ORGANIC
CHEMISTRY
GUIDE
to molecules

ORGANIC CHEMISTRY GUIDE
to molecules

MATEUSZ WOZNY

First published 2012

This book was set by the author and
printed and bound in the United States of America.

Karolinie i Malutkiej

Contents

Part V **Resonance & induction**

Part VI **Acid-base reactions**

http://www.ochguide.com

Part I STRUCTURE AND DRAWINGS

01 Introduction to the molecules

Atoms and ions constitute building blocks of the significant part of an inanimate world: metal ores and minerals, including gemstones and lava. Yet, chemistry is not so attached to elements as most people tend to think, because atoms are just one of the levels of structural complexity of matter. Many scientists are immune to their charms and find periodic table of elements a bore. Indeed, there is something more intriguing:

The best things are made of molecules...

...which are structures made of a *definite number of atoms interconnected in a specific way, through links we call* **chemical bonds**. Molecules are ubiquitous in Nature, because most atoms eagerly bond each other upon contact.

For us, atoms are the building blocks of molecules and nothing more. And we will need just a few: H hydrogen, C carbon, N nitrogen, O oxygen, and X **halogens**. The term "halogens" is a common name of four atoms lumped together because of very similar chemical behavior: F fluorine, Cl chlorine, Br bromine, and I iodine (by the way, P like phosphorous and S like sulfur, are also among VIPs, but in a basic organic chemistry course, only few molecules contain these two).

Hence, we have eight different atoms in our toolbox (H, C, N, O, F, Cl, Br, I), or just five if we treat halogens as one (H, C, N, O, X). A bit encouraging for an organic chemistry novice, right?

Anyway, these few atoms, bonding each other eagerly, are enough to generate the diversity of organic molecules – there are millions known to date. How is that possible?

The structure - not the atomic composition - makes a molecule unique

Look at the picture below, which shows a unique arrangement of exactly 10 C atoms, 20 H atoms and 1 O atom. This particular structure is a molecule chemists call **menthol**:

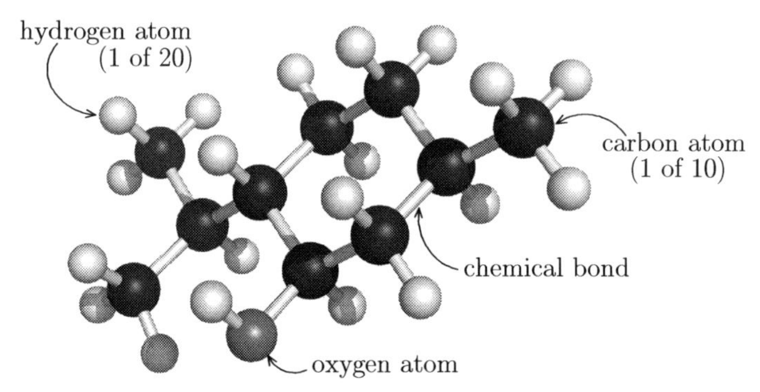

Fig. 1.1 Structure of a menthol molecule. Balls represent atoms, while sticks represent chemical bonds, which glue atoms together.

What makes a menthol molecule different from millions of others? The **structure** – a unique way in which the particular set of atoms is interconnected. The structure determines properties of a molecule and, therefore, it is much more significant, than the atomic composition.

We denote the composition with **chemical formulas**. For menthol, it looks like this: $C_{10}H_{20}O$, and simply describes which of the atoms, and in what number, make up the molecule. We do not need to write the subscript "1" on the right of O, because the very presence of the symbol denotes that *there is* an oxygen.

What if I gave you a pouch with 10 C atoms, 20 H atoms and O atom, and asked to glue them into some molecule? The number of contraptions you could make is more than enormous, and all of them would be completely different. Although the atomic composition is identical, structures differ substantially, and so do properties. We call *molecules, which have the same atomic composition but different structures* **isomers** of each other.

Below, I drew a **citronellol** molecule. Compare it with the one above. Menthol and citronellol are isomers of each other:

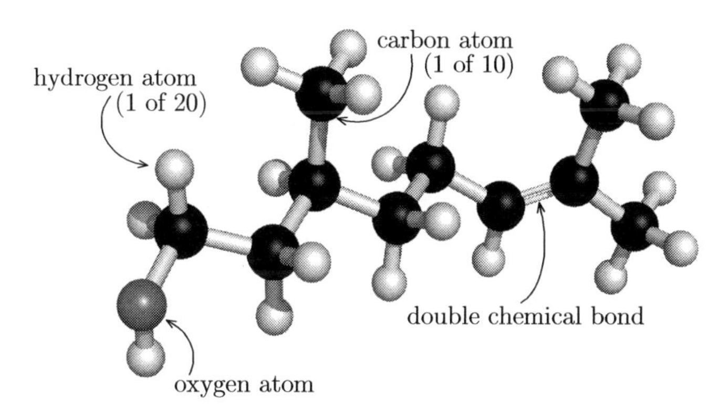

Fig. 1.2 Structure of a citronellol molecule. Citronellol is an isomer of menthol, because both are composed of 10 C, 20 H, and 1 O atom.

After high school chemistry, students tend to think of formulas, like $C_{10}H_{20}O$, as inseparable part of the chemistry course. This is no longer the case.

Look. The only information encoded within the simple formula is the atomic composition. That describes an immense number of isomeric molecules, which structures vary drastically. Therefore, organic chemistry lectures and textbooks are full of **structure drawings**, rather than formulas. As you will soon see, they are simplified versions of 3D pictures like the two above. And we should appreciate them, because determination of the true structure of some unknown molecule happens to be a difficult task. Simply, we cannot inspect it by the eye, because what our eyes can see are powders, crystals, or drips of...

The chemical compound

Menthol is present in peppermint, granting its mint smell and chilling property. Citronellol is an ingredient of rose oil. Properties of matter such as smell, color or state do not relate to molecules. They describe **chemical compounds**, because the compound is something we can see, smell or touch.

Compound is a *discernible heap of molecules of a particular structure.* A single molecule is so small that we cannot see it, even with the best optical microscope on the planet. Imagine that in a speck of a solid compound, or in a minute drip of a liquid one, the number of molecules is like hundreds of millions of millions of millions.

When such a vast number of menthol molecules, all identical clones, lie next to the other, we can see a small crystal. Only then, we experience properties of the compound: it is solid, colorless and has a mint smell.

On the other hand, citronellol molecules, jammed together, form a drip. As a compound, citronellol is a colorless liquid with a flowery scent. Physical features of menthol and citronellol vary because molecular structures of both compounds are dissimilar. Chemical properties are structure-dependent too, so chemical behavior of menthol and citronellol will also be different.

Chemists working in industry and academia need chemical compounds for their research, and buy them closed in bottles and jars from specialized international companies like Sigma Aldrich (U.S.) or ABCR (Germany).

Organic molecules are based on the carbon backbone

We perceive the world as 3-dimensional, and so its building blocks must be. Thus, all molecules are 3D objects. And all of them are based on the **carbon backbone.**

Just take a look at menthol and citronellol. In both cases, hydrogen atoms lard the backbone and one oxygen atom sits here or there. Indeed, C atoms are the basis of all organic molecules, and that is why some scientists call organic chemistry "the chemistry of carbon." Backbones are structural foundation of molecules, and as any foundation should be – they are remarkably stable. It means: inert to most chemical transformations. It is so, because C–C and C–H chemical bonds are strong.

What is a chemical bond?

As you might remember from high school chemistry, when two atoms collide, they may create a chemical bond.

Each of the bonds is made of one, two or three *pairs of electrons.* Why we talk about pairs? Because the process of bond formation is fair – one atom provides one of its electrons, the second atom does the same. A **bonding pair**, that is a **single chemical bond** forms.

Sometimes an atom donates two or three electrons at once, the second doubles the stake, and a **double** or a **triple chemical bond** arises. In each case, electrons used to build a bond escaped their atoms, and started a new life as bonding pairs lying in the common space between nuclei.

We depict the chemical bond by drawing strokes between both atoms. One stroke symbolizes the most popular single bond. Two strokes represent two bonding pairs of a double bond, like the one in the citronellol molecule, while a triple bond is drawn as three strokes (by the way, in organic chemistry, there are no bonds made of four pairs or more).

In general, *only few of all electrons a free atom contains, have the ability to form chemical bonds.* These are the only **chemically active electrons**, officially referred to as **valence electrons**.

The figure below depicts a cutting from the periodic table. It shows only these elements, which are salient here. Dots representing chemically active electrons surround the symbols:

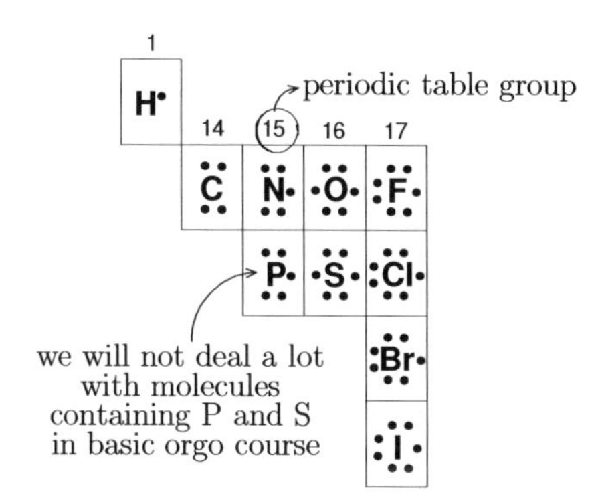

Fig. 1.3

Periodic table for a student of an organic chemistry. Number of chemically active electrons equals the rightmost digit of the number of a group where an atom in question lies.

Atoms and ions

Willy-nilly we went back to atoms. The end of the chapter is a good place to review them briefly (things from high school). Atoms are composed of three types of so-called **subatomic particles**:

electrons, e^-	carrying an electric charge of -1,
protons, p^+	with +1 charge, and
neutrons, n^0	with no charge at all.

Protons and neutrons pack themselves into an unimaginably tiny lump, which we call the **nucleus**. It always sits in the geometric center of an atom, while a relatively large sphere, formed by all electrons the atom has, surrounds it. Now, it becomes apparent that electrons have a leading role to play in chemistry, because they form chemical bonds, and therefore, are responsible for formation of molecules.

In all atoms, the number of positively charged protons and the number of negatively charged electrons are in perfect balance. It is not a coincidence. The balance guarantees that every atom, as a whole, is a neutral entity. For example, every C atom contains 12 e^- and 12 p^+, while H atoms, the simplest of all, are made of 1 e^- and 1 p^+ only. Of course, there are some neutrons in nuclei of all atoms except for H, but they are of no significance in chemistry. We ignore their very existence without remorse.

Anyway, the perfect p^+–e^- balance no longer holds, when we tear out an electron, or stuff an extra one into the atom. Such change in

the composition of an atom makes it a charged particle, which we call an **ion**.
When we remove one electron from a neutral atom, an ion with *the number of* p^+ *exceeding the number of* e^-, comes into being. It has an overall charge of +1 (e.g., H^+ or Na^+). We call positively charged ions **cations**.

On the other hand, when we stuff an extra electron into the atom, *the number of* e^- *exceeds* p^+, and we have a negatively charged ion, a.k.a. **anion** (e.g., H^- or Cl^-). Since many of the ions are even more stable than corresponding atoms, we treat them as equally important.

Next two chapters are a quick review of basic chemistry. Please do the problems carefully – they are your warm-up. Then, in Chapter 4, we start to delve into more interesting stuff.

02 Constructing molecules of C and H atoms

C atoms are the foundation of organic molecules, and Hs are their best friends. In fact, majority of organic molecules have backbones larded with H atoms, and likely but not necessarily sprinkled here and there with N, O or X. For now, we assume that Cs & Hs are all we have in our builder kit.

How C and H atoms form chemical bonds?

Ways in which atoms bond each other, depend on the number of chemically active electrons they have. Every H atom *always creates only one single bond*, because it has only one electron to share. It simply cannot do more. C atoms, on the other hand, have four chemically active electrons, and participate in formation of four bonding pairs. In addition, there are four different ways to do so, which we call **hybrids** (more officially "**hybridization** of C atoms"):

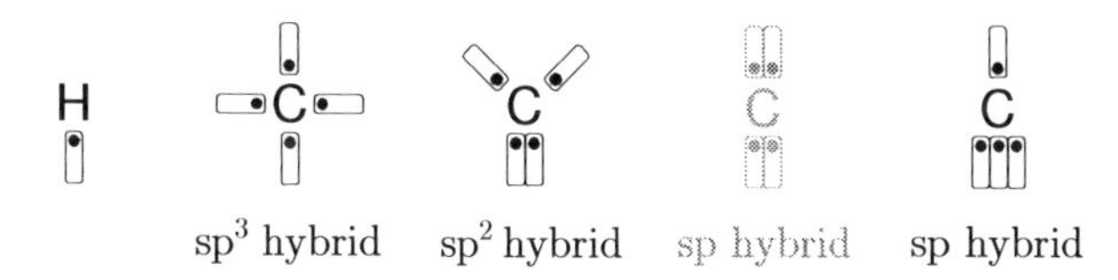

Fig. 2.1 Atomic bricks: H atom and four hybrids of C.

In molecules, C atom with 4 single bonds is the most common. Nevertheless, another possible hybrid is C with 2 single & 1 double bond, while C atom with 1 single & 1 triple is the last typical mode. The hybrid with 2 double bonds is also feasible, but this pattern of bonding is of interest for people tinkering with advanced chemistry. For us too rare to bother.

Now, let's construct some carbon backbones and cover them with Hs. Please, remember that we will always refer to *molecules made exclusively of hydrogen and carbon* as **hydrocarbons**. We make them with H & C exclusively.

The most banal backbones are chains of C sp^3 atoms

Write the letter C and assume it is the entire backbone. Not at all intricate, right? It is the simplest backbone ever.

To draw a realistic hydrocarbon molecule we have to fulfill the bonding desire of the C atom. In order to do that, four H atoms are needed. The result is the CH_4 molecule, a main ingredient of natural gas. The name of this simplest hydrocarbon molecule is **methane**:

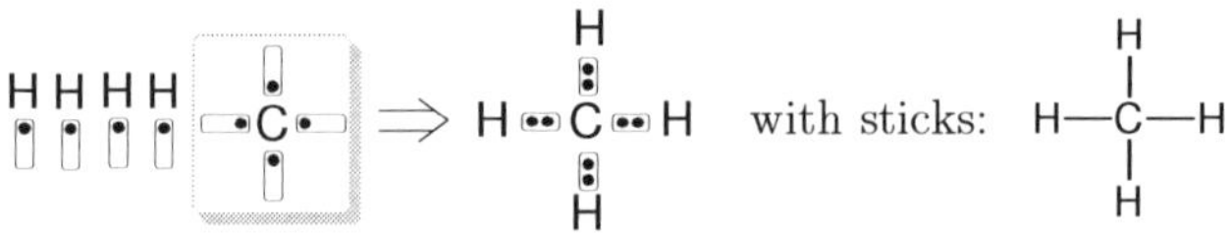

Fig. 2.2

Try a backbone with 2 C atoms bonded via single C–C bond. How many H atoms do we need now to make a real hydrocarbon? There is a place for six. Let's attach them to get **ethane**:

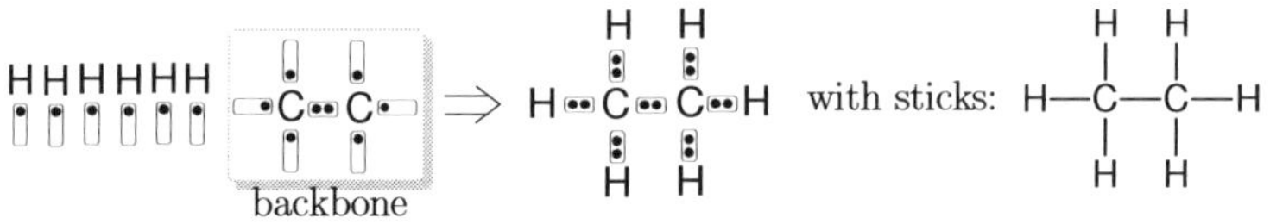

Fig. 2.3

Repeat the process to obtain longer chainlike hydrocarbons. A **propane** molecule is based on 3 C backbone. With 4 C chain you get **butane**, with 5 **pentane**, 6 **hexane**, 7 **heptane**, 8 **octane**, 9 **nonane**, and with 10 C chain, larded with as much as 22 H atoms, you get a hydrocarbon called **decane**.

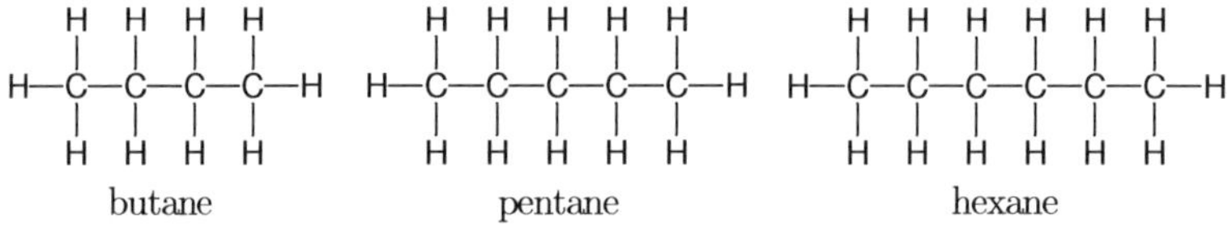

Fig. 2.4 Structures of three other chainlike hydrocarbons.

The sequence does not stop here, but for us, it is enough. Now you had better beware: you should learn names of all these molecules, from methane to decane, by heart. Treat it as if it was counting from 1 to 10 in some foreign language.

C sp^2 and C sp hybrids: double and triple bonds

We can stuff double C=C and triple C≡C bonds to our backbones. These are just two other possible ways to glue Cs. No mystery. Draw 2 C backbone, like in ethane, but this time join Cs with a double or a triple bond. Add necessary H atoms thereafter.

The simplest hydrocarbon with C=C bond is **ethene** (or **ethylene**), while the second one with C≡C bond is **ethyne** (a.k.a. **acetylene**):

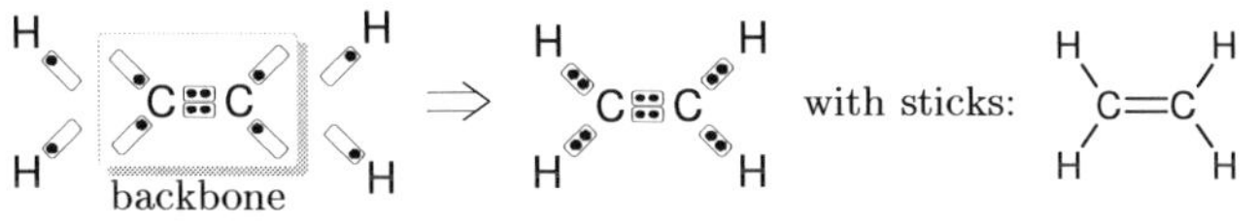

backbone ⇒ with sticks: H₂C=CH₂

Fig. 2.5 Ethene (ethylene).

H—C≡C—H backbone ⇒ with sticks: H—C≡C—H

Fig. 2.6 Ethyne (acetylene).

Making cyclic and branched backbones

It would be foolish if all backbones were chainlike arrangements of C atoms joined with C–C, C=C or C≡C bonds. Indeed, menthol and citronellol molecules had more intricate backbones. This is because we can increase the structural diversity of our contraptions, by making them **branched** or **cyclic**:

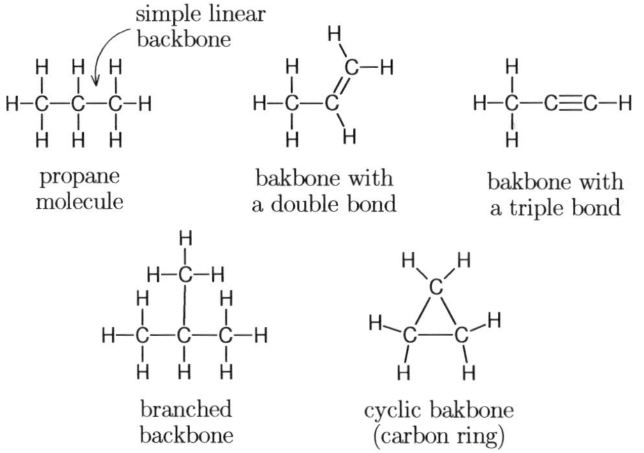

simple linear backbone

propane molecule

bakbone with a double bond

bakbone with a triple bond

branched backbone

cyclic bakbone (carbon ring)

Fig. 2.7

Branches and **carbon rings** are far more the rule, than the exception. In addition, there is nothing wrong in combining all ideas into one molecule – feel free to build backbones containing double and triple bonds, branches, and rings at once.

Indeed, the number of backbones to concoct is enormous. That is why structures of organic molecules are so diverse. The only thing you must remember is that we *never draw C atom with more than 4 bonds*! Nature strictly forbids it.

Problems

Draw hydrocarbons, which backbones are just like described:

2.1 The backbone is made of 5 C atoms: 4 C in a chain, the second has a 1 C branch attached.
2.2 The backbone is made of a 4 C ring, and there is a 2 C branch.
2.3 There is a C≡C bond and one other C atom.
2.4 The backbone is a ring, has a total of 5 C atoms and there are two C=C bonds.
2.5 The backbone of this chainlike molecule is made of 3 C atoms, and there is one double bond.
2.6 Interesting molecule with backbone made of 5 C atoms, one of which has four bonds to four carbons.
2.7 The backbone is a chain made of 6 C atoms, and there are three double bonds (do not use =C= hybrid!).
2.8 The backbone is made of a 3 C ring, and there is a 1 C branch.
2.9 The backbone is a 3 C ring, two Cs bond to 1 C branches.
2.10 Backbone is made of a 3 C ring, one C bonds two 1 C branches.

2.11 Are there any isomers among molecules **2.1-2.10**?

Feel free to rotate, flip and flop your drawings

Are all of your drawings (problems **2.1-2.10**), identical to mine (answers **2.1-2.10**)? I bet they are not. Are you wrong? Not necessarily. That is because any drawing can be freely rotated, flipped or flopped. It does not change the structure of the molecule, just as you do not transmogrify into your neighbor, when you toss from side to side during sleep. Convince yourself that all drawings below depict the same molecule:

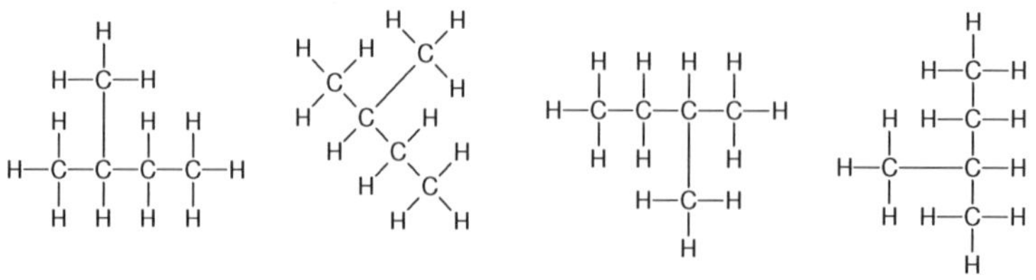

Fig. 2.8 Clockwise rotation of the drawing.

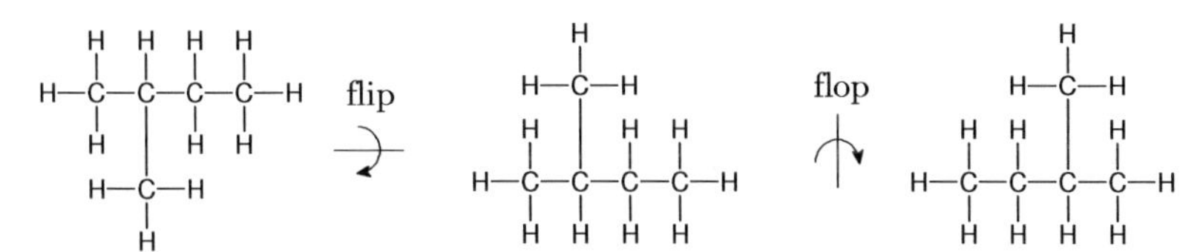

Fig. 2.9 Flipping and flopping of the drawing.

03 Molecules of C, H, N, O and X atoms

N, O and X (halogens) atoms have so-called **lone pairs** of electrons. "Lone" means that atoms do not share them to form bonds in normal situations – they do not socialize.

N atom has 5 chemically active electrons, but 2 of them form a lone pair. N uses 3 remaining electrons to form chemical bonds. Following patterns of bonding are possible: 3 single bonds, or 1 single & 1 double bond, or 1 triple bond.

O atom has 6 chemically active electrons, including two lone pairs. 2 remaining electrons wait to participate in bonds formation, and there are two possible ways to do so – the hybrid with 2 single bonds, or the hybrid with 1 double bond only.

Finally, every halogen atom has 7 chemically active electrons, including as much as three lone pairs. Only 1 electron is keen to form a bond, so Xs form one bond just like H:

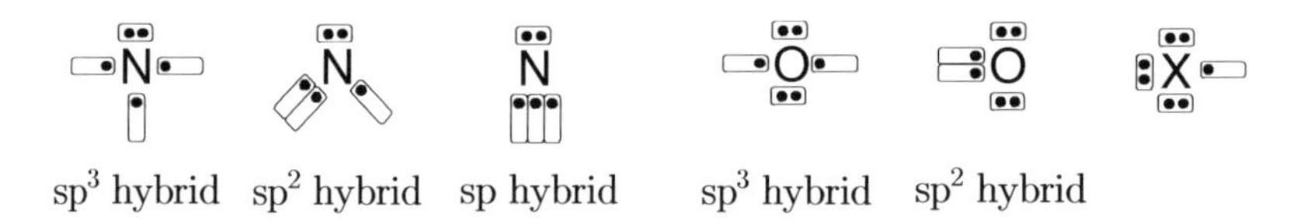

Fig. 3.1 Atomic bricks: X atoms, hybrids of N and hybrids of O.

Any molecule, which contains N, O or X atom, is no longer a hydrocarbon. Just like menthol or citronellol, because both of them have O within the structure. So let's try to make a citronellol molecule. Follow me, while I am translating an eye-appealing ball-stick picture from Chapter 1, into the structure drawing.

On Fig. 1.2, we see five C atoms arranged into 4 C chain with 1 C branch. Easy to do. Then, there are two C atoms joined with a double bond – C sp^2 hybrids for sure, and two more C sp^3 hybrids. The figure below shows all 10 C bricks, though I have already glued up the first fragment:

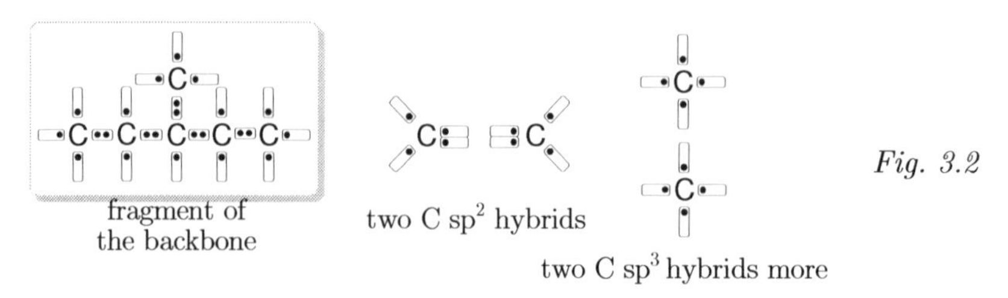

Fig. 3.2

We can easily create remaining C=C and C–C bonds, to finish the backbone. It is going to become citronellol in a minute:

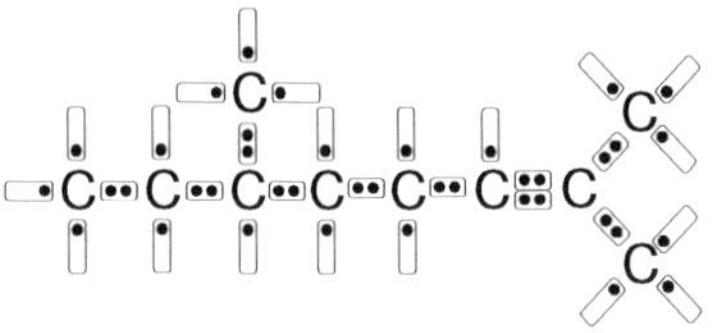

Fig. 3.3

According to Fig. 1.2, we should attach O atom to the leftmost C of the contraption. This O has two single bonds, so it must be O sp^3 hybrid. Mind two lone pairs:

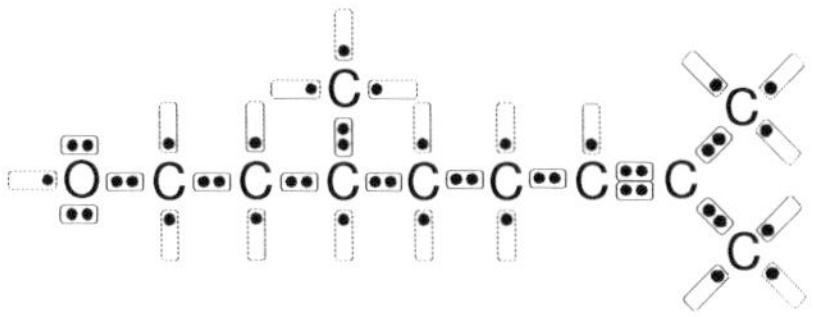

Fig. 3.4

We have 10 of 10 C and 1 of 1 O bonded correctly, but the "thing" above *is not yet a molecule.* Nothing like this could exist in reality, because atoms have not yet fulfilled their bonding desire. What is still to be done, is larding them all with 20 H atoms:

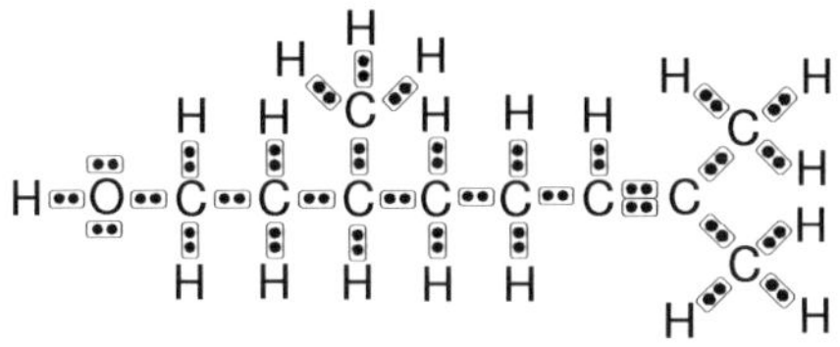

Fig. 3.5

All these dots are a bit too gaudy, so I redrew the picture with sticks. Mind, that lone pairs remain as pairs of dots, so that they stand out from the set of bonding pairs:

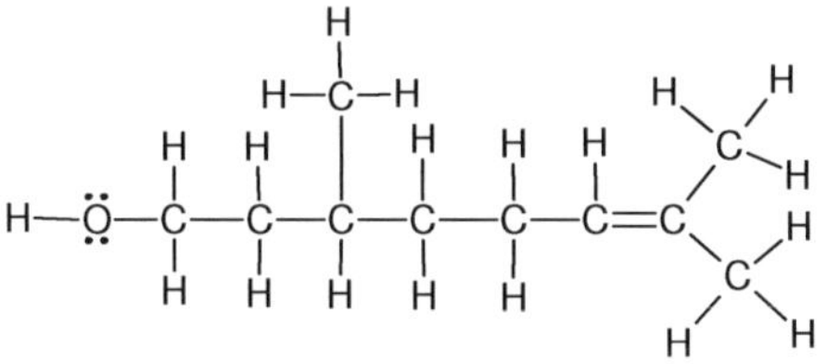

Fig. 3.6

In organic chemistry, we prefer sticks, because they are not so time-consuming as dots. Nevertheless, it would be even faster to draw a *condensed drawing* with all H atoms adjacent to C atoms they bond:

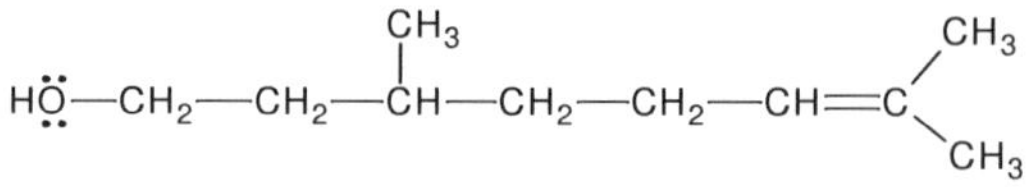

Fig. 3.7

This is more readable. Compare the result with the original 3D picture of citronellol. It is clear that both represent the same interconnections between the $C_{10}H_{20}O$ set of atoms. We can say that both pictures show *the same structure*, and therefore, depict *the same molecule*.

The only thing that differs is *the shape*. Carbon chain in the 3D picture looks like a zigzag. Well, chains are zigzags in real. We will incorporate this feature into our drawings in the next chapter, where I will show you how to draw in a professional manner.

Problems

3.1 Redraw drawings of citronellol (Fig. 3.5 and 3.6) as zigzags.
3.2 By analogy, draw normal and condensed drawings of menthol.

N, O and X atoms *attached* to the carbon backbone

N, O and X (halogens) atoms are usually attached to the carbon backbone, just like O in menthol and citronellol. No mystery in that, but remember: since C, N, and O atoms can form multiple bonds, we can attach them not only via a single bond. For example, O atom can form a double bond just like in the example below:

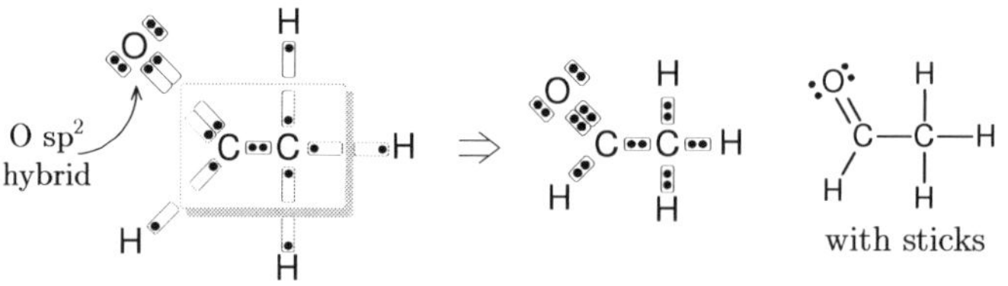

O sp^2 hybrid

with sticks

Fig. 3.8

We will often refer to some fragments of the molecule as **substituents**. Substituent is *anything other than H* attached to... *something else*. For example, O atom in the molecule above is a substituent of the first C atom.

Now, try to construct molecules with N atom as a substituent. It can be attached to the backbone via single C–N bond, a double C=N bond, or even a triple C≡N bond. Try halogens – they are much simpler to handle, because there is no fuss with multiple bonds. Xs always form only a single bonding pair.

Problems

Use sticks to draw structures of molecules according to descriptions below (fill a remaining valence with H atoms):

3.3 1 C backbone, O attached to it via single bond.
3.4 1 C backbone, N attached to it via single bond.
3.5 1 C backbone, Br attached to it via single bond.
3.6 1 C backbone, O attached to it via double bond.
3.7 1 C backbone, N attached to it via double bond.
3.8 1 C backbone, three Cls attached to it.
3.9 1 C backbone, F and Cl attached to it.
3.10 2 C backbone, N attached via single bond.
3.11 2 C backbone, N attached via triple bond.
3.12 The ring of 4 C, O attached via double bond.
3.13 The ring of 5 C, N attached via single bond.
3.14 The ring of 5 C, O atom and 1 C branch attached simultaneously to one of C atoms of the ring.

3.15 Are there any isomers among molecules **3.3-3.14**?

Read the following descriptions, and draw matching molecules (more than one in each case):

3.16 3 C in a chain, O attached to the first C
3.17 3 C in a chain, O attached to the middle C
3.18 2 C backbone, Br attached to one C, and N to the other

3.19 Compare problems **3.16** and **3.17**. Why there would be no sense to ask for "3C in a chain, O attached to the third C?"

N and O atoms *inserted* in the carbon backbone

Since N and O form more than one bond, they can join more than one C at once. It will look like *inserting them into the backbone*. Below I drew such a molecule. It contains O atom, which bonds two Cs at once:

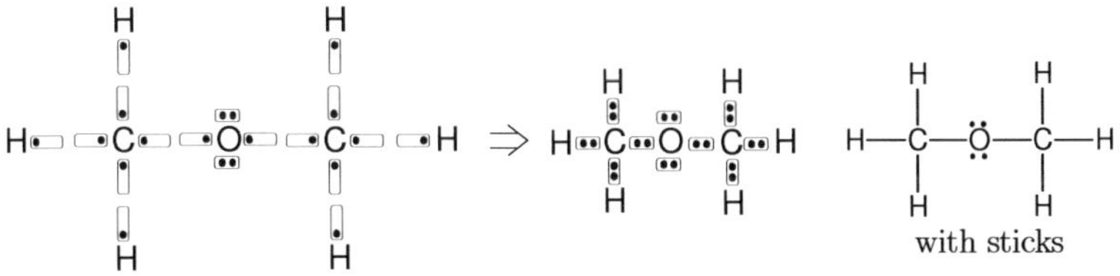

H H H H H H
H C O C H ⇒ H C O C H H—C—O—C—H
H H H H H H
with sticks

Fig. 3.9

It is also possible for an atom like N or O to be part of the carbon ring. Why not? It does not violate any of the bonding patterns:

with sticks

Fig. 3.10

Problems

Use sticks to draw structures of molecules according to descriptions below (*hint:* acyclic means "not cyclic").

3.20 An acyclic molecule made of 2 C and one N; the N bonds to both of Cs via single bonds.
3.21 3 C in an acyclic molecule, N bonds *all* of them via single bonds.
3.22 4 Cs and 1 O in a ring, single bonds only.
3.23 4 Cs and 1 O in a ring, and there are two C=C bonds.
3.24 4 Cs and 1 N in a ring, and there are two C=C bonds.
3.25 Ring of 5 Cs and 1 N, single bonds only; N atom additionally bonds 1 C (atomic composition of the molecule is $C_6H_{13}N$).

Octet rule justifies bonding patterns of all atoms

In this and previous chapter, we used several atomic bricks to construct examples of organic molecules. And there is hugeness of possible structures. The common feature is that *atoms in organic molecules form fixed number of bonding pairs.* A carbon atom always has four – we say its **valence** is 4. H and X atoms always have only one bond, O exactly two, while the valence of N is 3. How can we explain it?

Nature wants atoms to *have 8 electrons around.* This is a so-called **octet rule**. And because free atoms have less than 8, they feel an urge to bond. Bonding increases the number of electrons around.

Look. If you share one of your electrons to form a bonding pair with another atom, you end up with an overall number of electrons increased by one, because the bonding pair is a common property. Both entrants benefit, and atoms form as many bonds as needed to reach the octet:

C atoms	have 4 e^-, lack 4 e^- must form 4 bonding pairs to achieve the octet
N atoms	have 5 e^-, lack 3 e^- must form 3 bonding pairs (2 e^- remain as a lone pair)
O atoms	have 6 e^-, lack 2 e^- form 2 bonding pairs (4 e^- remain as lone pairs)
X atoms	have 7 e^-, lack only 1 form 1 bonding pair (three lone pairs left)
H atoms	create only 1 bonding pair, because have only 1 e^- to share; they never fulfill the octet and never complain

Bonded atoms may have less than 8 electrons, but such arrangement is very unstable and rare. On the other hand, having more than 8 is completely impossible. This would be a *violation of the octet rule.*

04 How to draw a molecule professionally

When computer software renders a 3D picture of a molecule, it looks clear thanks to shades and imposed illusory perspective. But we are unable to do such things with paper and pencil. Hand drawings are messy, and unfortunately time-consuming. Since learning organic chemistry requires drawing many molecules, we need a faster method, so that you can save your time.

Get rid of H & C atoms, as well as lone pairs

Way to optimize drawings is intuitive and results from simple reasoning. Look at molecules you have drawn already – there are so many H symbols that they dominate images. It would always be the case. Getting rid of Hs means saving a lot of time.

Therefore, the first agreement is as follows: *never draw H atoms attached to the carbon backbone.* Analogically, since C is a fundamental, all-pervasive atomic brick, the second agreement is to *never write the letter C.* Let the sticks meet in a point instead. Lastly, we *omit dots representing lone pairs* on N, O and X atoms. And this is how three agreements make drawing of the menthol molecule simplified:

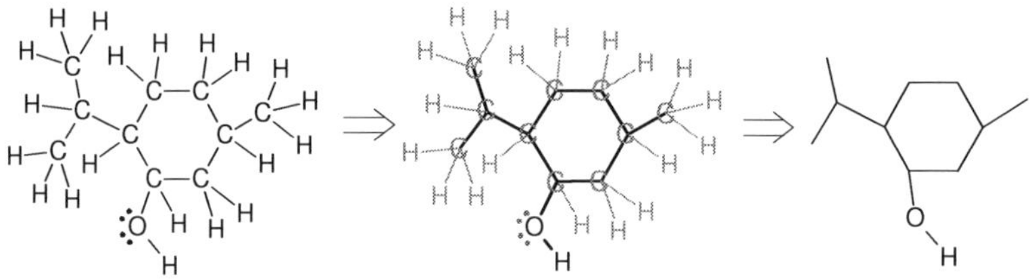

Fig. 4.1 On professional structure drawings, we hide all C symbols, lone pairs, and H atoms attached to Cs.

Looking at professional structure drawing, remember that H atoms are still there (just not drawn over and over); C atoms lie in every intersection and on each end of a stick, while N, O and X atoms still carry their lone pairs.

We can now try to translate a bit messy drawing of the citronellol molecule into a professional one. However, be careful! – a disaster happens when you simply remove Hs, Cs and lone pairs, leaving the chain as a straight line:

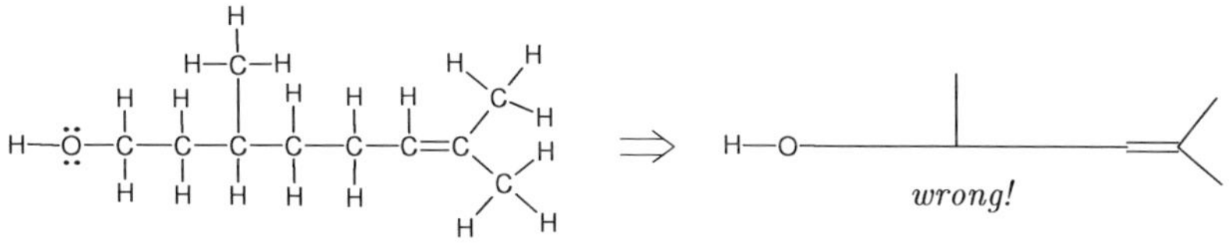

wrong!

Fig. 4.2

I wrote at the end of the last chapter, that carbon chains are zigzags in reality. Now it is clear that we certainly need to draw them as such, because otherwise the positions of hidden Cs are not visible:

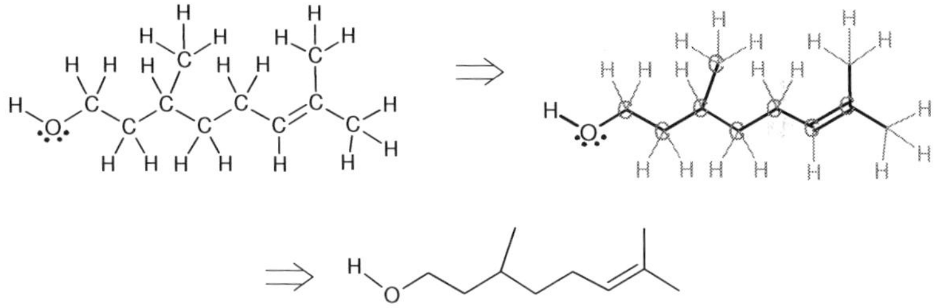

Fig. 4.3

When you are a beginner, it is very easy to make a mistake – lots of students happen to draw chains with one C too many. To avoid that you should count Cs, while drawing the zigzag. But remember that the first stroke already represents two C atoms, so we start counting from "two." Each new stick is an extra C.

Good structure drawing fits an imaginary honeycomb

Both structures of menthol and citronellol are close to perfection, but they look a bit kicked. So, here comes another hint: good structure drawing fits the honeycomb, which makes it more symmetrical:

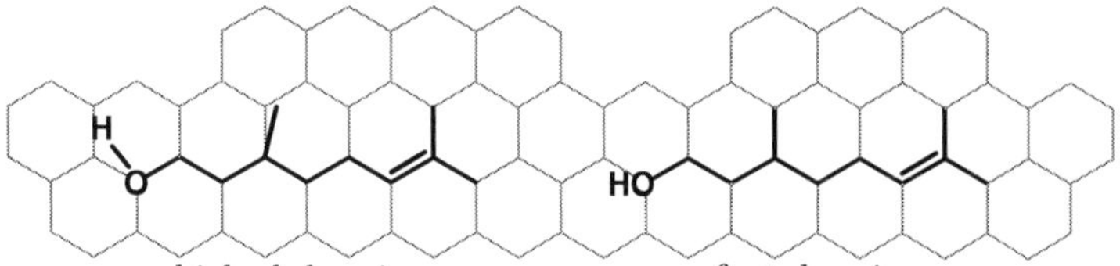

kicked drawing

perfect drawing
(note: it's good to hide the O-H bond)

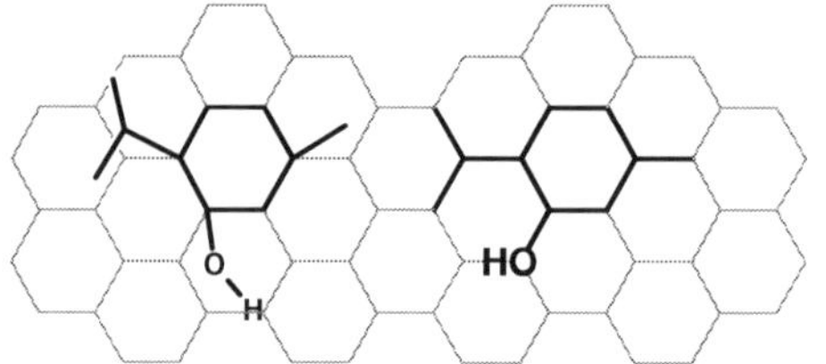

kicked drawing perfect drawing

Fig. 4.4

Mistakes to avoid

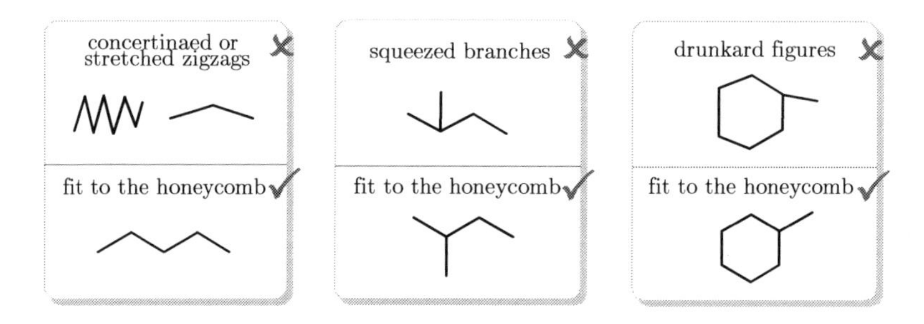

We draw carbon rings as regular figures – keep equal angles and bond lengths. Moreover, retain the symmetry when you draw a bond sprouting from the ring:

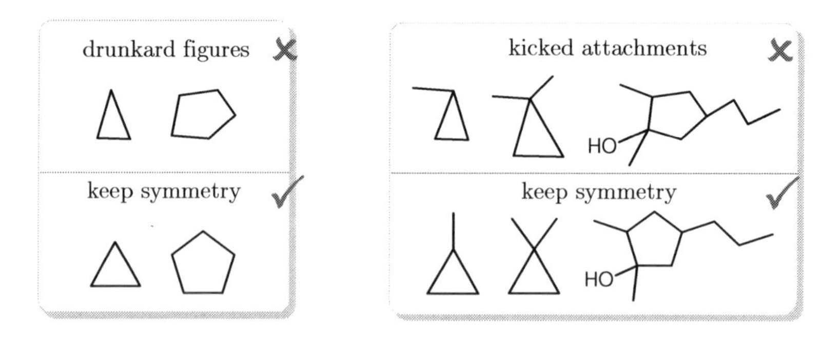

Rings other than six-membered will not fit the honeycomb. The same happens to the triple C≡C bond. Remember: we never bend it! Draw both substituents sprouting from C≡C in a linear fashion, because it is like that in reality:

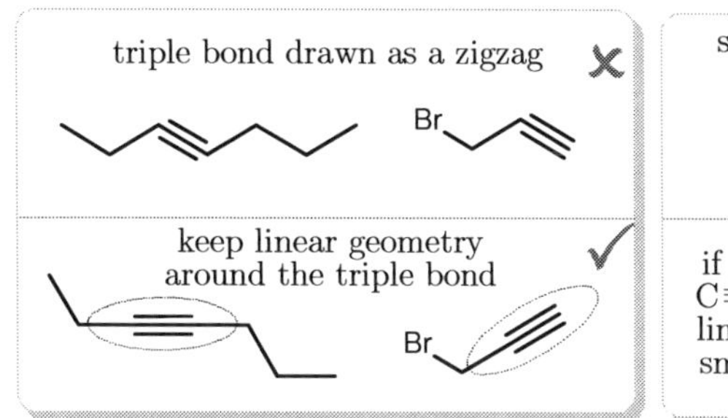

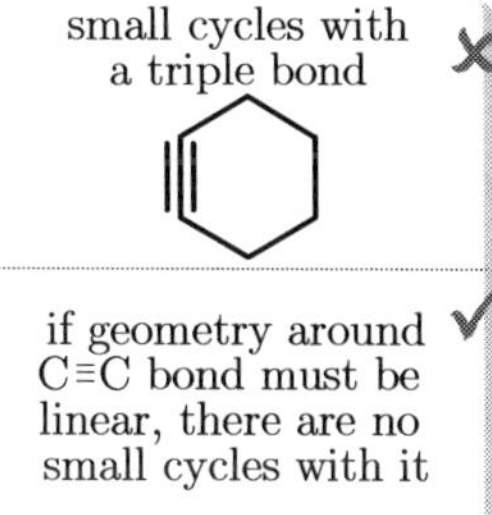

Problems

Draw following chain hydrocarbons:

4.1 ethane **4.2** propane **4.3** butane
4.4 pentane **4.5** hexane **4.6** heptane
4.7 octane **4.8** nonane **4.9** decane

By the way, methane CH_4, drawn in a professional way, would be a single point. We simply rewrite "CH_4" formula instead.

Now draw following cyclic hydrocarbons. Note that their names start with a characteristic "cyclo" chunk:

4.10 cyclopropane **4.11** cyclobutane **4.12** cyclopentane
4.13 cyclohexane **4.14** cycloheptane **4.15** cyclooctane

Following molecules have already appeared in previous problems. Redraw them in a professional way:

4.16 **4.17** 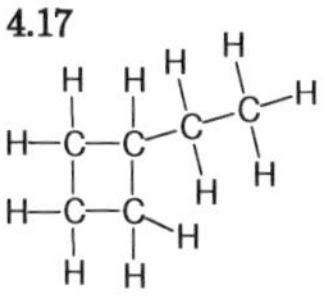 **4.18** 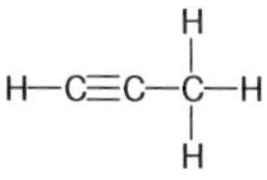 **4.19** 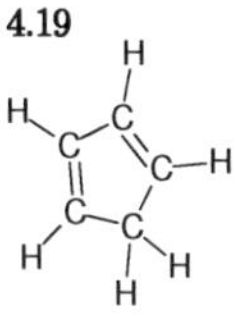 **4.20**

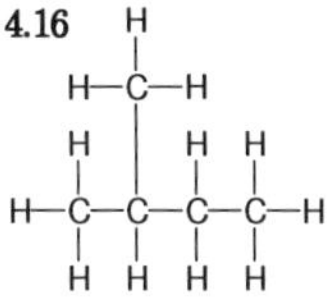

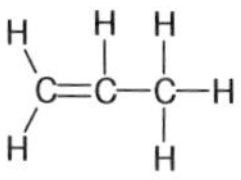

4.21 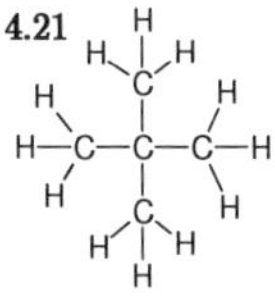 **4.22** 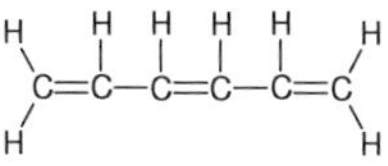 **4.23** 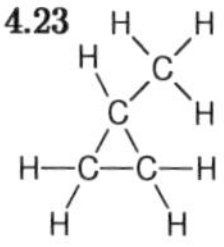 **4.24** 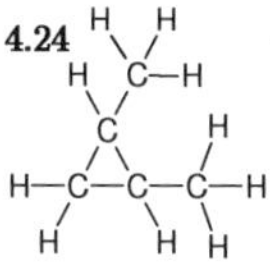 **4.25**

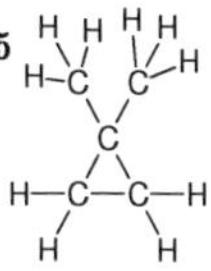

4.26 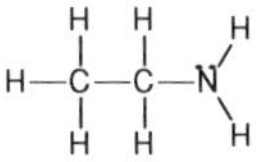 **4.27** 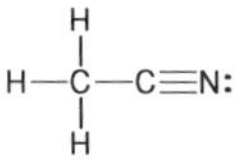 **4.28** 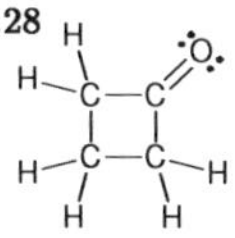 **4.29** 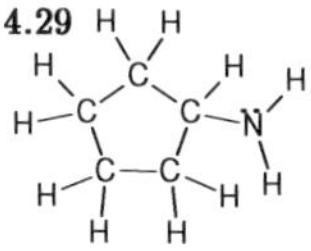 **4.30**

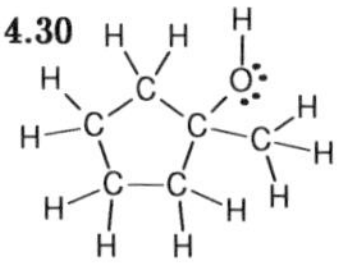

4.31 4.32 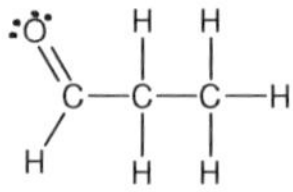 4.33 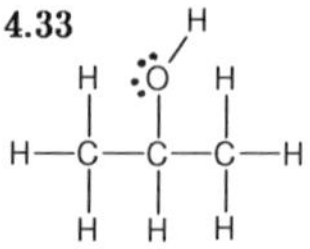 4.34

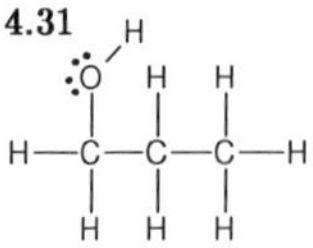

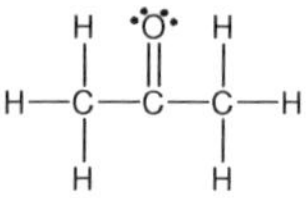

4.35 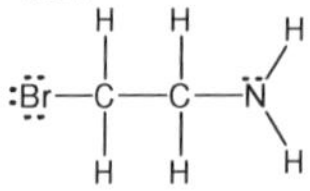 4.36 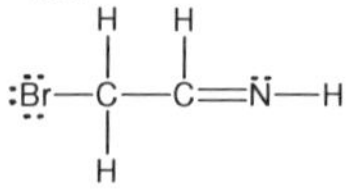 4.37 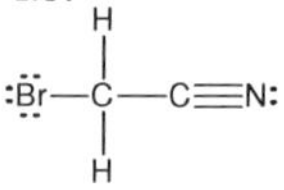 4.38

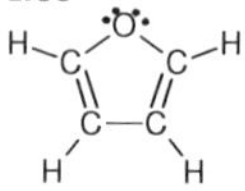

4.39 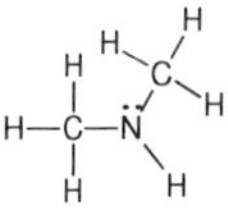 4.40 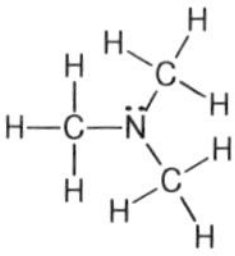 4.41 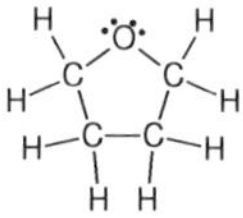 4.42 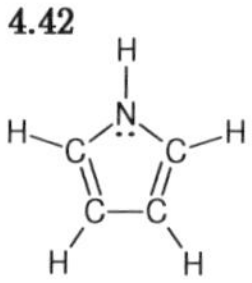 4.43

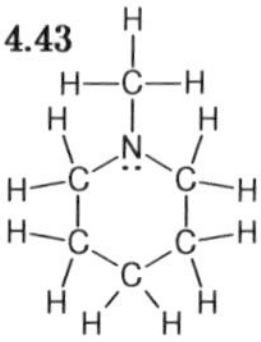

Try to do the same with some new molecules:

4.44 4.45 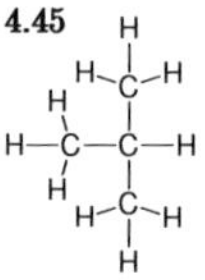 4.46 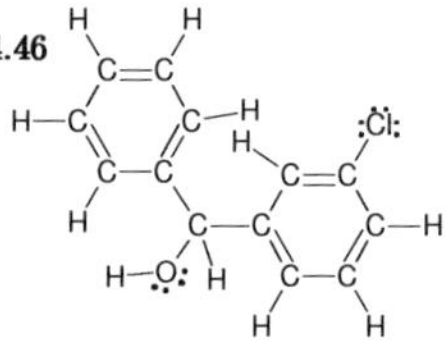 4.47

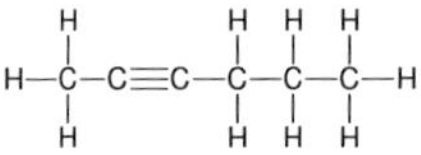

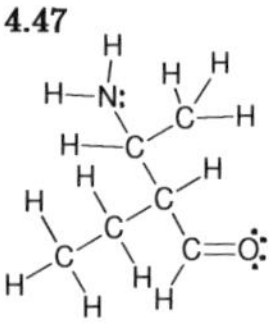

4.48 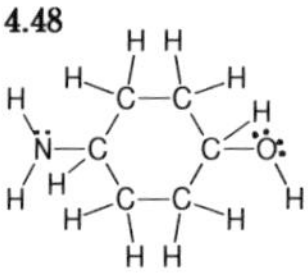 4.49 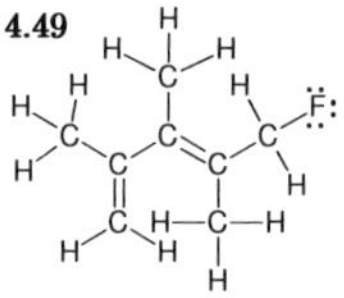 4.50 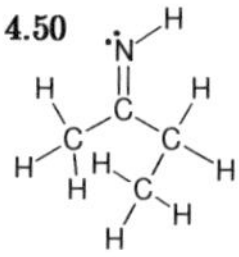 4.51

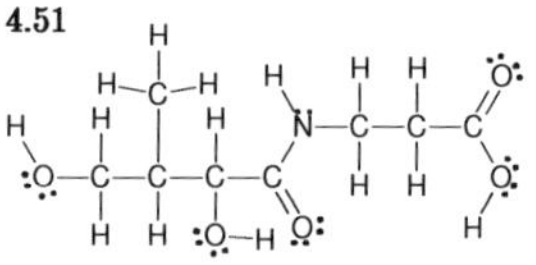

Reading drawings relies on locating every hidden C or H atom (it becomes automatic as you gain experience). In the following problems mark every point, which symbolizes C atom, and assign the number of H atoms attached:

example

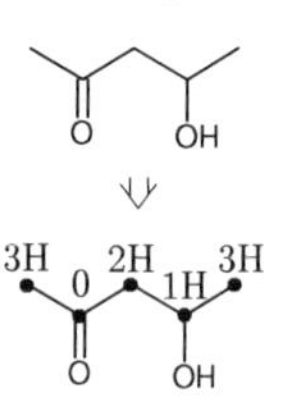

4.52 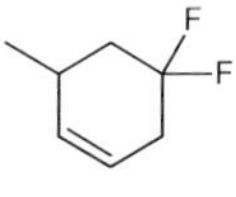 4.53 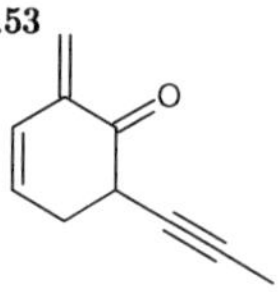 4.54 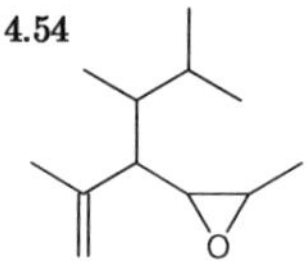 4.55  4.56

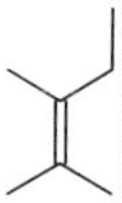

4.57 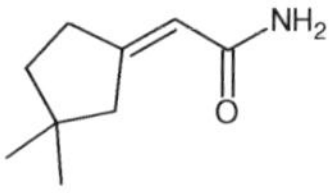 4.58 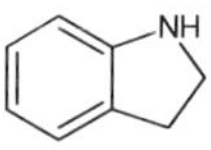 4.59 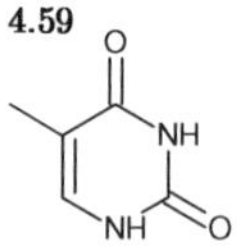 4.60

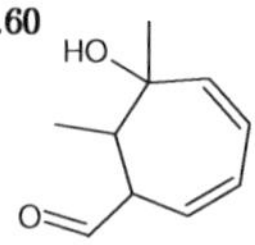

Rotating, flipping, flopping and bending chains

Professional drawings put emphasis on the backbone of the molecule. Only N, O and X atoms are shown explicitly with their symbols, and the rest is a planar map of connections within the carbon backbone. Therefore, the drawing is clear, and detailed only as necessary. All chemists in the world draw molecules this way.

As you know, we can rotate, flip and flop drawings, because it does not change the structure. Imagine a group of friends, who gather around a map of the city they are visiting. Each of them looks at it from a different direction, but still, there is no doubt, they look at the plan of the same city.

Look at the figure below, where I drew a simple hydrocarbon several times. 7 C atoms make up the backbone of this molecule, and there are 16 Hs undrawn. Note that we can distinguish a 1 C branch from 6 C chain. We will refer to such a fundamental part of the backbone, as the **root backbone**, from which smaller branches sprout. In other words, the carbon backbone = root backbone + branches. Anyway, all these drawings are correct representations of the same molecule:

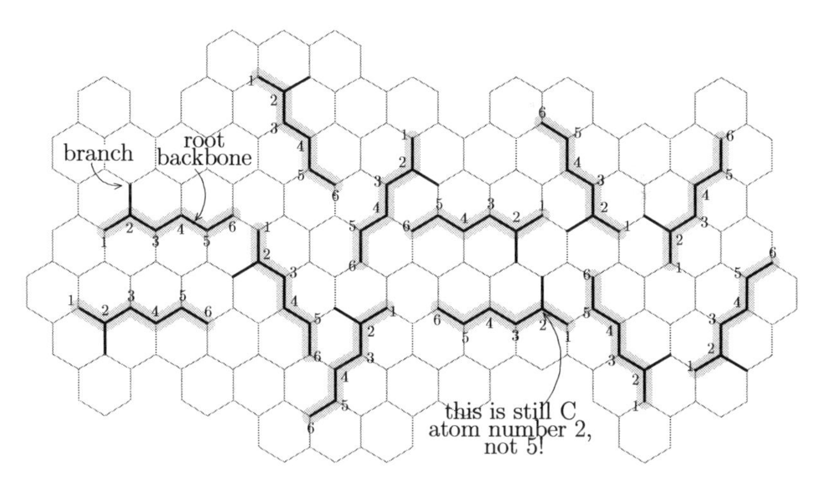

Fig. 4.5

I have numbered Cs in the root backbone intentionally. We usually do that in order to communicate easily – what if someone asks you: "At which C of the root backbone is the branch located?" It is clear that the branch is at 2nd C of the root, although it feigns 5th sometimes.

When you need to check whether several drawings depict the same molecule, test the identity by rotating, flipping and flopping them. However, there is yet another thing we can do to professional drawings, and it makes comparisons a bit more difficult.

We are allowed to bend chains. It must be so, because otherwise the only way to draw really long molecules would be to do it on a toilet paper. And *there are* such molecules out there. For example, some plastics are made of chains as long as tens of thousands of C atoms.

Look at the picture below. It illustrates what I mean. Still, all these drawings should fit with the honeycomb:

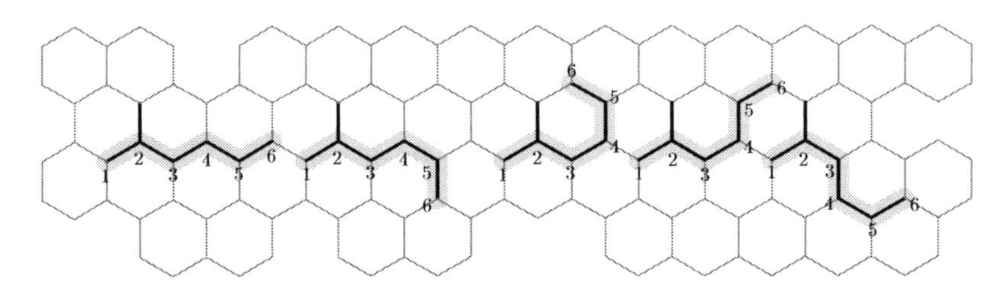

Fig. 4.6 The same hydrocarbon – bonds bending this time.

05 Free software as a study aid

Molecule editors are programs for drawing chemical structures. In general, they are designed for professional chemists, but it may be useful to have one during learning organic chemistry.

Popular and full-fledged molecule editors are **ChemDraw**, **ChemSketch** and **MarvinSketch**. First one is obscenely expensive. For you, ChemSketch or MarvinSketch is probably the best choice, since there are freeware versions for educational use. Useful web applet called **Chemicalize** is also worth recommendation:

ChemSketch	www.acdlabs.com/resources/freeware/chemsketch/
MarvinSketch	www.chemaxon.com/download/marvin/for-end-users/
Chemicalize	www.chemicalize.org (click "draw" button)

Learning how to draw molecules in the editor is not so difficult, and there are lots of online tutorials, including videos on YouTube.

What are the benefits? Editors assist you during drawing, helping you to avoid structural mistakes (editor instantly marks atoms with an improper valence). They also provide a structure cleanup, which makes perfect drawings from messy ones. There is also an option to make all H atoms attached to the carbon backbone explicitly shown, and to generate simple chemical formulas in a second.

Other features are useful for further parts of the curse. Firstly, stereo structure view is a mode designed to watch molecules in three dimensions, which is better than gluing models from plastic balls and sticks (Part III). Secondly, there is a generator of IUPAC names (Part II). You should be careful with this, however, because programs are not ideal in this regard, and mistakes happen from time to time. Lastly, editors can accurately assign *R*, *S*, *E*, *Z*, *cis*, and *trans* descriptors (Part IV).

I recommend downloading and trying molecule editors especially to those of you who study at the faculties of chemistry. Eventually you will have to use it anyway.

Part II NOMENCLATURE

06 Functional groups & classes of compounds

The variety of molecules you have seen so far, might raise the feeling, that organic chemistry is a mess. It is time to tidy things up; we need some sort of classification.

In organic chemistry, we group molecules into **classes**, characterized by common structural moieties, which we refer to as **functional groups**. It will later become apparent that functional groups have a profound role to play. *They are chemically reactive parts of molecules and impose characteristic chemical properties.* Recall that carbon backbones are remarkably stable. Therefore, what defines chemical nature of a molecule must be indeed associated with other things attached or stuffed within.

Both functional groups and classes of compounds have names, so there is a bunch of new words to learn. Fortunately, during basic organic chemistry, we are going to familiarize with only the most important ones, and these are not too many.

Superclass of hydrocarbons

As you already know, hydrocarbons are compounds, which molecules are made of H and C atoms exclusively. Among hydrocarbon superclass, we distinguish several classes, drastically differing in properties.

Alkanes are hydrocarbons in which Cs are bonded only via single C–C bonds. C–C and C–H bonds are very strong, and in consequence, alkanes are chemically stable and exceptionally unreactive:

Fig. 6.1 Alkanes

The word **cycloalkanes** is used in reference to alkanes containing carbon rings, like the two in the above picture. Another useful term, which chemists use constantly, is an **alkyl group**. Alkyl groups are any chains/branches of alkane-type, that is containing C–C bonds only.

Alkenes are hydrocarbons with at least one double C=C bond per molecule. By analogy, **cycloalkenes** are compounds with a carbon ring containing C=C bond. And finally, we refer to hydrocarbons with one or more C≡C bond as **alkynes**.

Both C=C and C≡C bonds are functional groups, because they are chemically reactive, in their own ways. We call C=C functional group "carbon-carbon double bond" or **alkenyl group**, while C≡C is a "carbon-carbon triple bond" or **alkynyl group**.

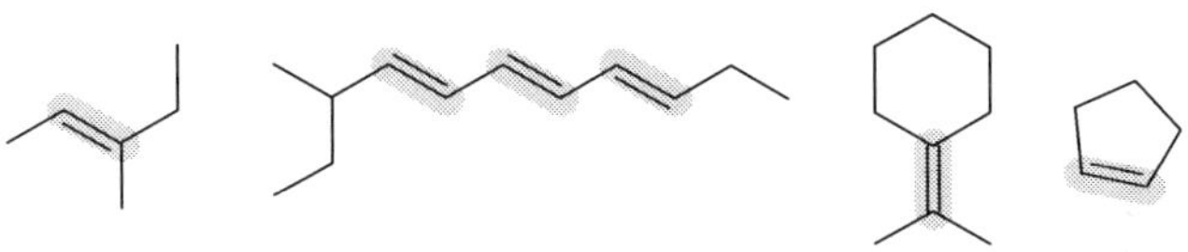

Fig. 6.2 Alkenes

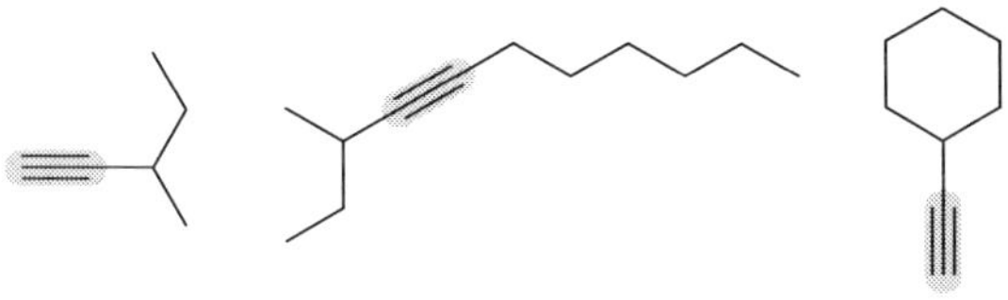

Fig. 6.3 Alkynes

Last but not least, **arenes** are any hydrocarbons, which molecules contain characteristic structural motifs, called **aromatic rings**. The most prominent representative is **benzene ring**, which is a 6 C ring, drawn with three C=C bonds inside. The corresponding molecule – **benzene** – is the simplest of all arenes:

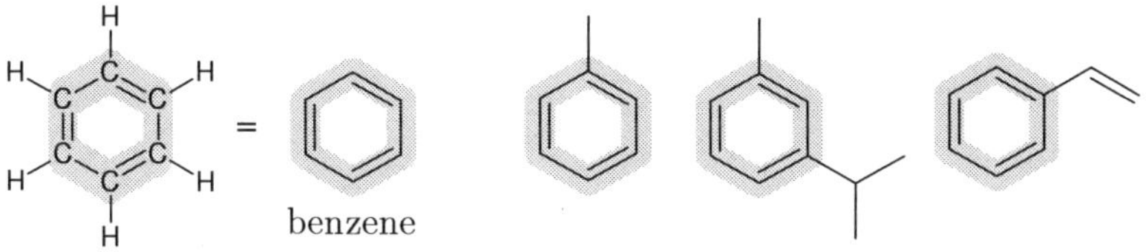

benzene

Fig. 6.4 Hydrocarbons belonging to the class of arenes, with a benzene molecule as the simplest representative of the class.

The benzene ring, as a whole, has specific chemical properties. We treat it as a unique functional group. Its role in chemistry is significant, and organic textbooks devote entire chapters to molecules containing benzene ring in their structures. Importantly, it does not behave similarly to alkenes with C=C bonds. That is why arenes and alkenes are completely different classes.

Alcohols and ethers

Alcohols are a class of organic compounds, which molecules contain an OH functional group, called a **hydroxyl group**, directly attached to the backbone. We saw it in familiar menthol and citronellol molecules. Ethanol, CH_3CH_2OH, an ingredient of alcoholic beverages, is another example. All alcohols share similar chemical properties arising from the chemical nature of the hydroxyl group.

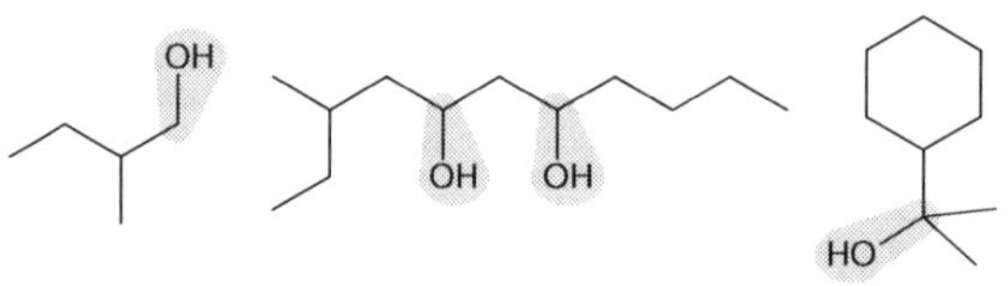

Fig. 6.5
Alcohols

An oxygen atom can also bond two C atoms at once, and then it becomes a new functional group, because the properties are significantly changed. We call it an **ether group**, and the class of compounds is **ethers**. All ethers share similar chemical properties arising from the chemical nature of an ether group, and – in the same time – they are very distinctive from alcohols:

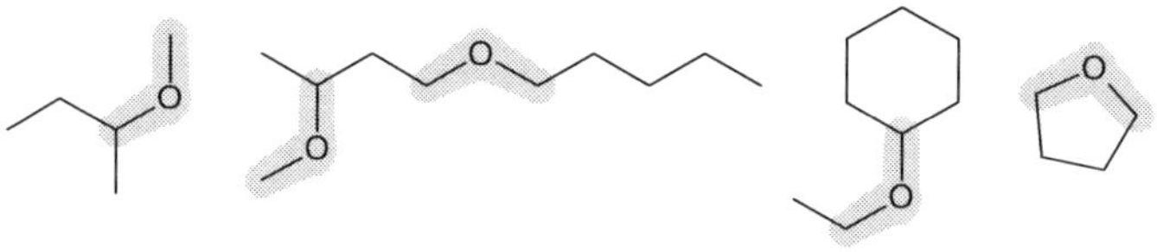

Fig. 6.6
Ethers

Ketones and aldehydes

An oxygen atom can also be attached to the backbone with a double C=O bond. This is yet another functional group, called a **carbonyl group**. Its presence characterizes two classes of compounds: **aldehydes** and **ketones**. Aldehydes are molecules with the carbonyl group at the end of the carbon chain, while ketones have it inside the chain or directly on the ring:

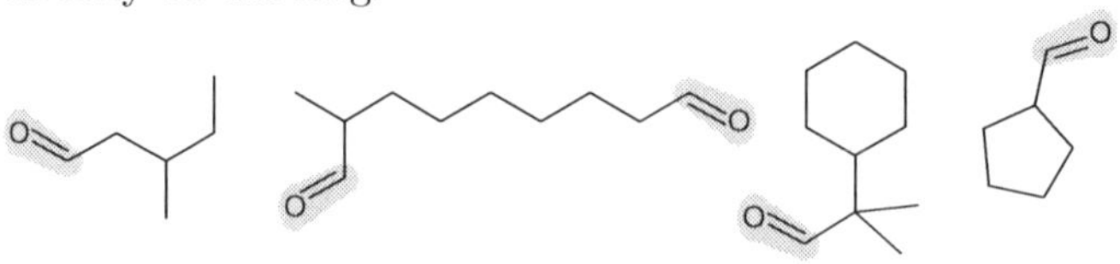

Fig. 6.7
Aldhydes

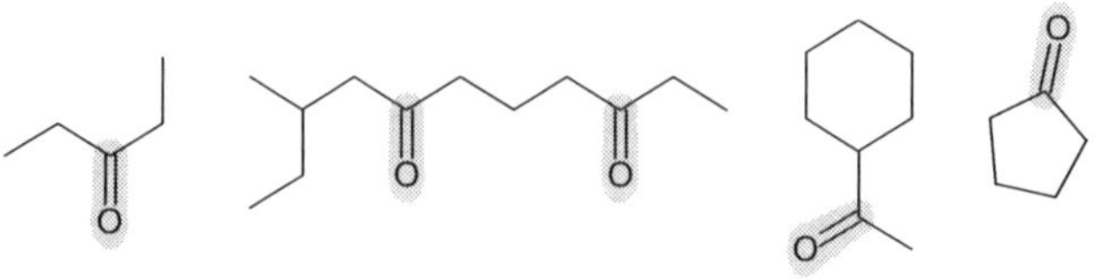

Fig. 6.8
Ketones

Amines - primary, secondary and tertiary

We can attach a N atom to the backbone via one single bond, giving rise to the NH_2 functional group. However, N can also be bonded via single bonds to two or three Cs at once. We call all these **amine functional groups**, and it defines the class of **amines**:

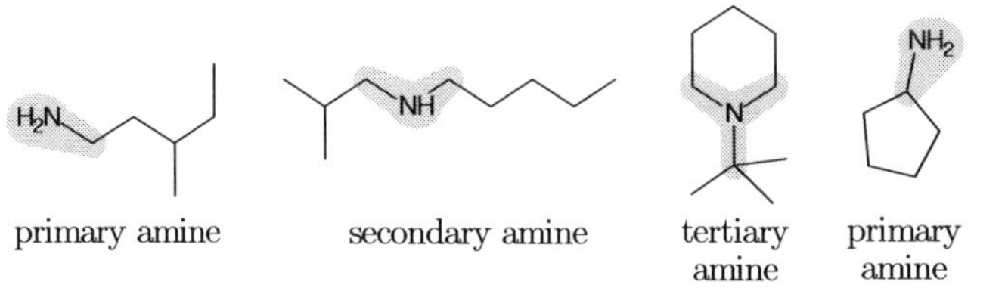

primary amine · secondary amine · tertiary amine · primary amine

Fig. 6.9 Amines

We differentiate amines into **primary** (N bonds to one C), **secondary** (N bonds to two Cs), and **tertiary amines** (N bonds to three Cs), but it is only to *emphasize the difference in structure.* Note that the relation between primary amine and secondary amine, is like a relation between alcohol and ether. Nevertheless, we do not classify various amines as entirely different classes, because their chemical properties are very similar.

Imines and nitriles

When you attach N to the backbone via a double bond, you get a new functional group called an **imine group**. Imines can be primary, or secondary, depending on how many C neighbors an N atom has:

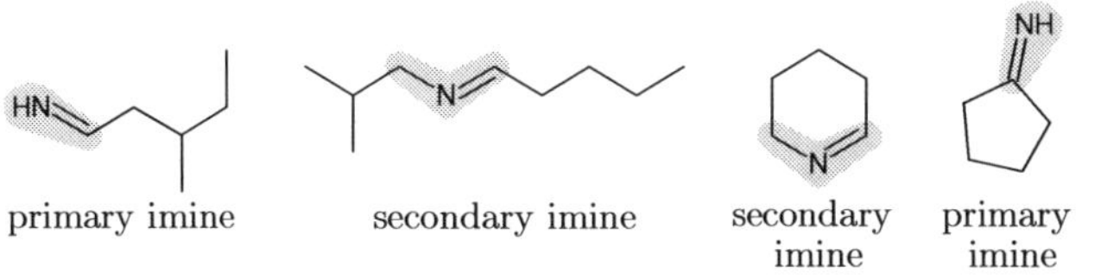

primary imine · secondary imine · secondary imine · primary imine

Fig. 6.10 Imines

On the other hand N atom attached to the backbone via a triple bond is a functional group called **nitrile group**. Neither imines nor nitriles are especially prominent, so you will not spend much time studying them during basic course:

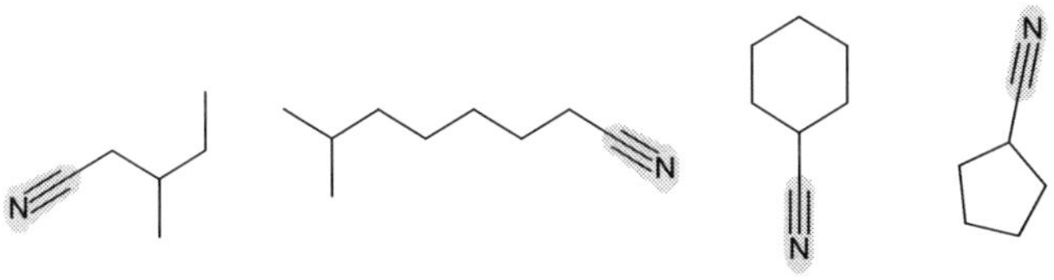

Fig. 6.11 Nitriles

Halogenated compounds

Any halogen atom in a molecule is also a functional group, because X is another type of a reactive site, which imposes specific chemical properties on a molecule. Halogen atoms are characteristic of the class of **halogenated compounds**:

F Cl I Cl Cl Cl Br

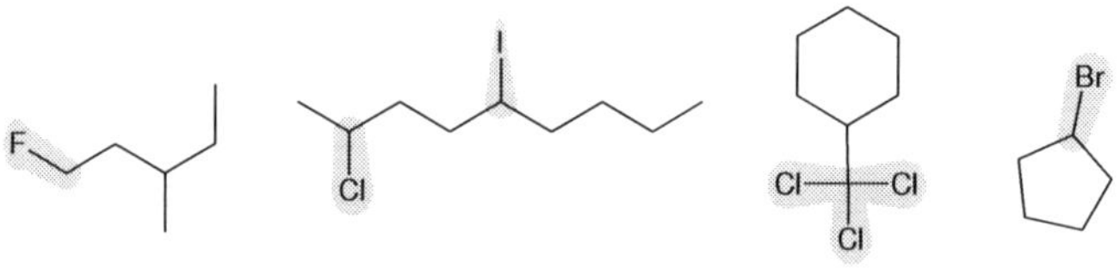

Fig. 6.12 Halogenated compounds

Superclass of carbonyl compounds

Recall that carbonyl functional group is an O atom attached to the backbone via double bond C=O. We meet this group in aldehydes and ketones, but that is not the end, because the carbonyl group is a part of some larger functional groups too.

The superclass of carbonyl compounds contains aldehydes and ketones, but also carboxylic acids, amides, acyl chlorides and esters. A **carboxylic acid** functional group is COOH:

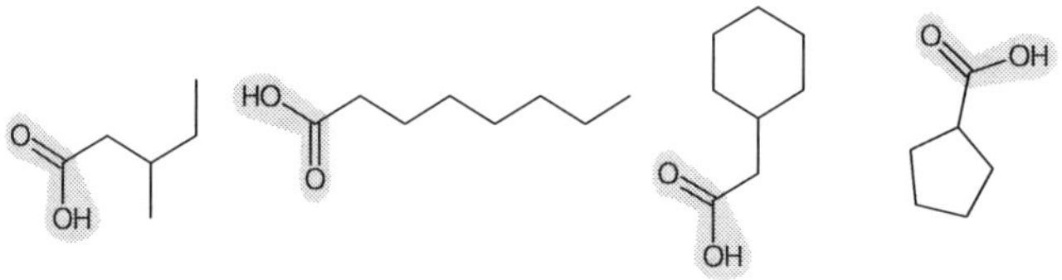

O OH HO O O OH O OH

Fig. 6.13 Carboxylic acids

Amide functional group is a combination of C=O with N. Just like in case of amines, we can divide amides into primary, secondary and tertiary depending on the number of C atoms to which N atom bonds:

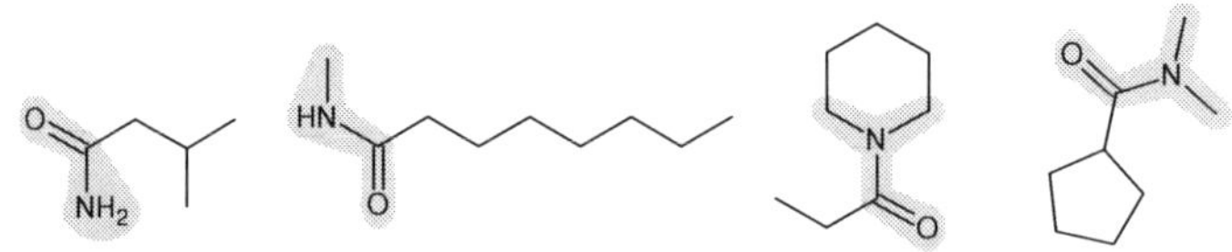

O NH_2 HN O N O O N

primary amide secondary amide two different tertiary amides

Fig. 6.14 Amides

Acyl chlorides have a functional group, which is a combination of C=O with Cl atom:

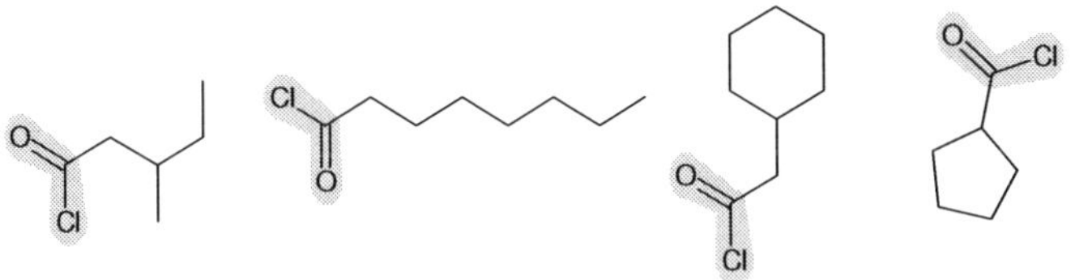

Fig. 6.15 Acyl chlorides

Finally, **esters** have a functional group similar to COOH, but with O sp^3 atom bonding C instead of a H atom:

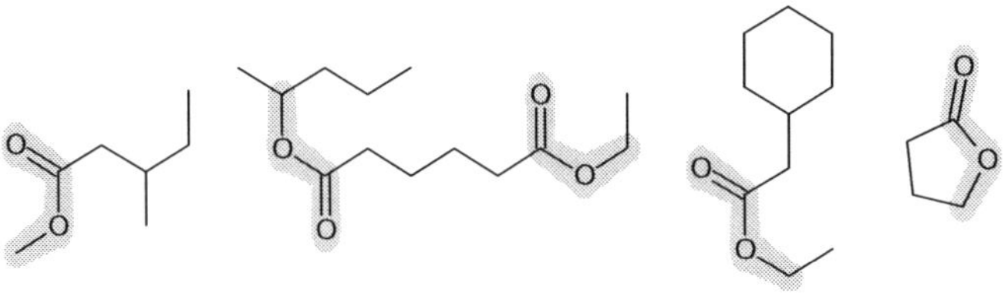

Fig. 6.16 Esters

Problems

Determine the class of organic compounds each of the following molecules belongs to. Encircle and name every functional group:

6.1 6.2 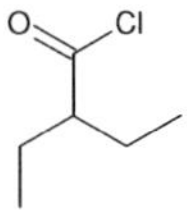 6.3 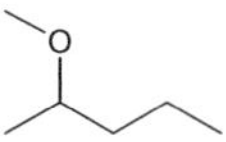 6.4 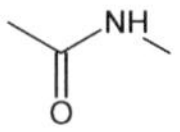 6.5

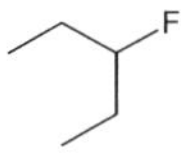

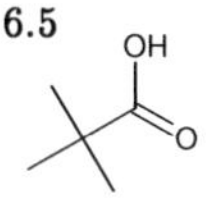

6.6 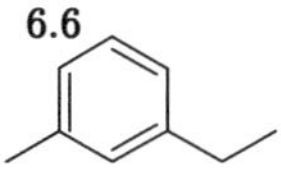 6.7  6.8 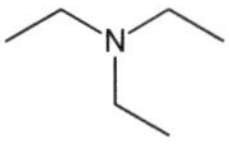 6.9 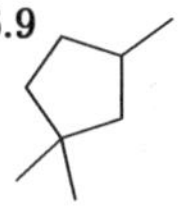 6.10

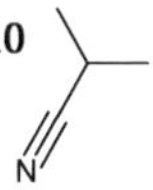

6.11 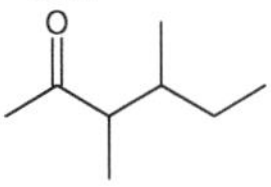 6.12 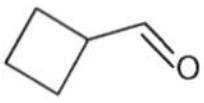 6.13 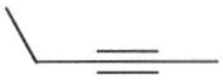 6.14  6.15

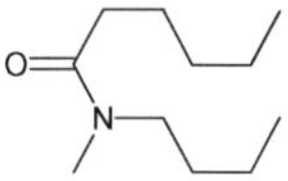

6.16 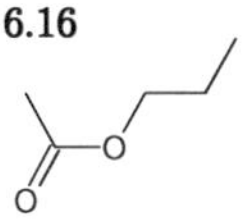 6.17 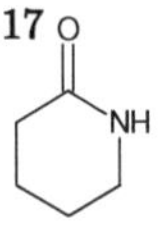 6.18 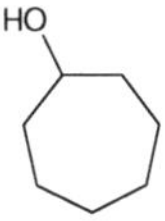 6.19 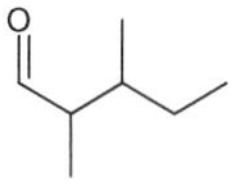 6.20

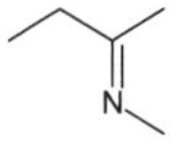

6.21 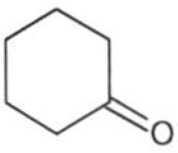 6.22 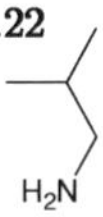 6.23 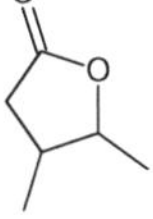 6.24 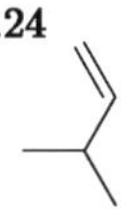 6.25 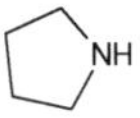 6.26

Superclass of aromatic compounds

Aromatic compounds are an important superclass, which describes all compounds with molecules having an aromatic ring in the structure.

Arenes are simultaneously aromatic compounds and hydrocarbons. However, it is not always the case, and many aromatic compounds are not hydrocarbons in the same time. For example, aspirin isn't made of H and C atoms exclusively:

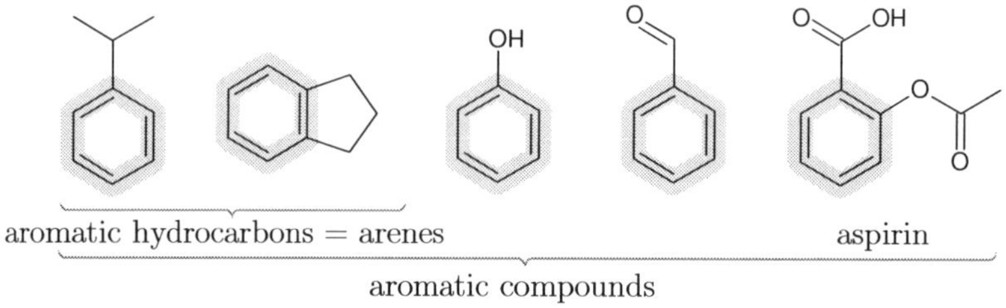

OH
O
O OH
O
O
aromatic hydrocarbons = arenes
aspirin
aromatic compounds

Fig. 6.17 Aromatic compounds contain the aromatic ring (shaded).

Aspirin and other examples from above have more than one functional group per molecule. Because of the benzene ring, they will behave like other aromatic compounds. Nevertheless, presence of other structural motifs – hydroxyl, aldehyde, carboxylic acid and ester groups – makes their chemical properties richer.

Molecules with two or more different functional groups are common in Nature

In fact, most organic molecules have more than one functional group. It substantially enriches their chemical properties, and final behavior is a combination of properties of all functional groups. It is correct to classify aspirin as aromatic compound, as well as carboxylic acid or an ester.

However, there is an official version of classification, according to which we assign such molecules to classes following *the table of priority of functional groups.* For instance, when a molecule contains OH and NH_2 groups, we classify it primarily as an alcohol, because OH wins in priority with NH_2:

priority	functional group	class of compounds
1	carboxylic group	carboxylic acids
2	acid chloride group	acid chlorides
3	ester group	esters
4	amide group	amides
5	nitrile group	nitriles
6	aldehyde gr. a.k.a. formyl gr.	aldehydes
7	carbonyl group	ketones
8	hydroxyl group	alcohols
9	amine group	amines
10	imine group	imines

11	ether group	ethers
12	C=C bond a.k.a. alkenyl gr.	alkenes
13	C≡C bond a.k.a. alkynyl gr.	alkynes
14	aromatic ring	aromatic compounds
15	halogen atom	halogenated compound
16	alkyl group	alkanes

Another nuance is that when a particular functional group is *directly attached to the benzene ring*, we can add the word "aromatic" before the name of the class. For example, we primarily classify aspirin as carboxylic acid, but it is even better to refer to it as aromatic carboxylic acid.

Problems

Encircle and name all functional groups in the structures below. Then classify molecules using the table of priority of functional groups. Remember to add the word "aromatic" when the highest priority functional group is directly attached to the benzene ring:

6.27 **6.28** 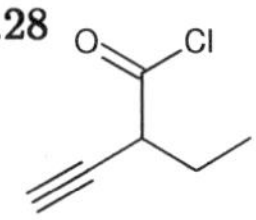 **6.29** 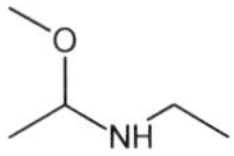 **6.30**

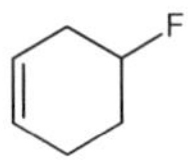

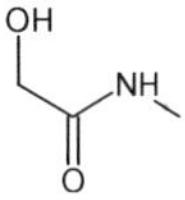

6.31 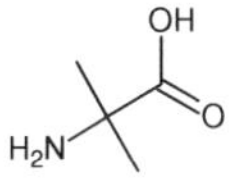 **6.32** 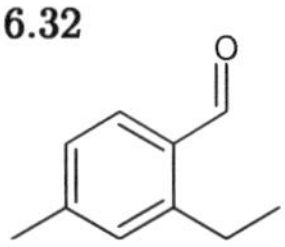 **6.33** 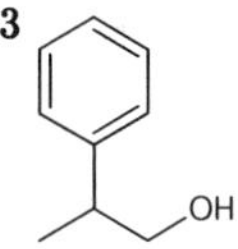 **6.34**

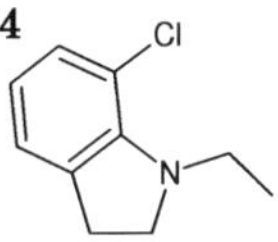

6.35 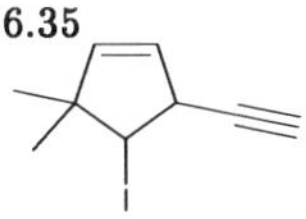 **6.36** 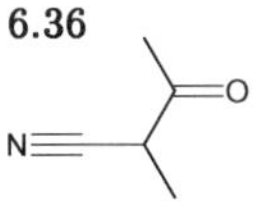 **6.37** 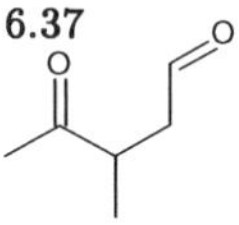 **6.38**

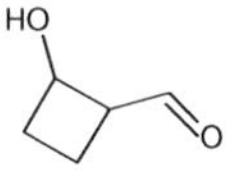

6.39 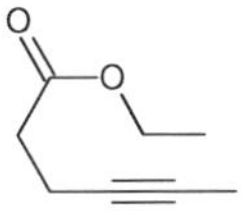 **6.40** 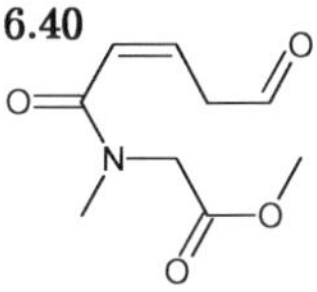 **6.41** 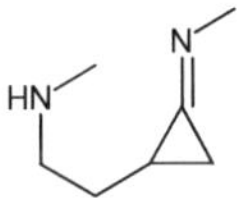 **6.42**

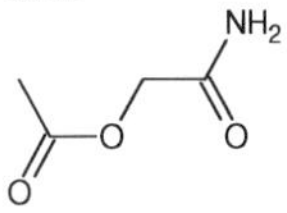

6.43 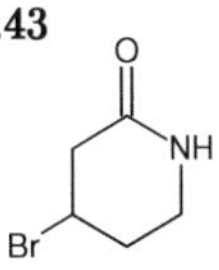 **6.44** 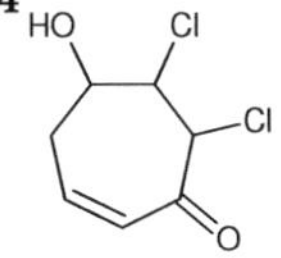 **6.45** 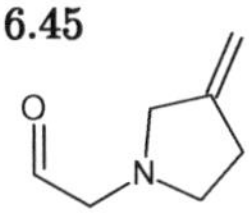 **6.46**

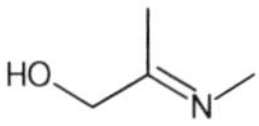

6.47 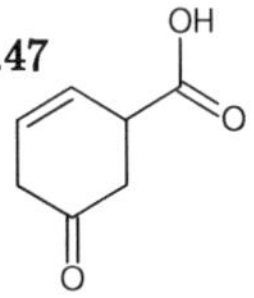 **6.48** 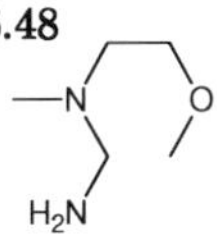 **6.49** 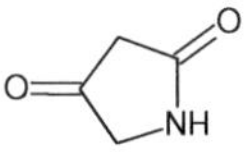 **6.50**

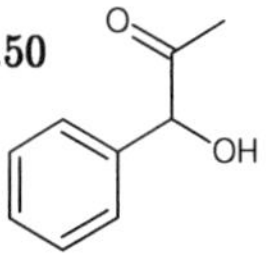

6.51 F F O F Cl F F

6.52 HO Br

6.53 HO O H_2N

6.54 O O HN

6.55 How many H atoms are there in the molecule **6.51**?

Molecules with more than one functional group of the same type

Chemists will refer to an ester with two ester groups as a **di**ester. They will call a nitrile with three nitrile groups a **tri**nitrile. It is easy – we simply add following multiplying "chunks":

2	3	4	5	6	7	8	9	10
di	**tri**	**tetra**	**penta**	**hexa**	**hepta**	**octa**	**nona**	**deca**

Problems

Classify following molecules using multiplying "chunks":

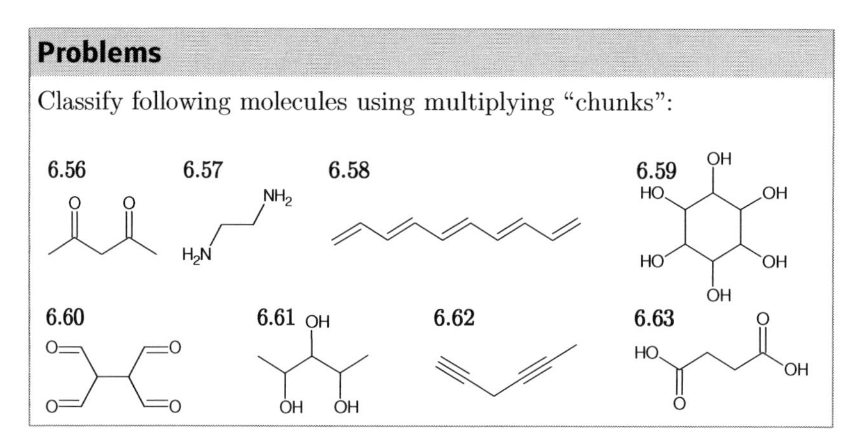

Summary

Functional groups are reactive parts, which impose specific chemical properties on a molecule. Hence, classification of organic compounds according to the type of functional group is a reasonable way to tidy up the variety of organic structures. We put them into proper drawers labeled with names like amines, esters, or amides. Most organic chemistry textbooks teach organic reactions by dividing them into groups characteristic for certain classes.

I recommend learning names of functional groups and classes of organic compounds by heart. However, you can also treat this chapter as a reference for the future.

07 "Chunks" of the systematic name

Name of a molecule is its unique ID. Many molecules with some historical background or significance in science and technology have nice one/two word names, called **trivial names** or **common names**. Menthol and citronellol are two examples we have already met. Others are aspirin, natural gas, glucose, caffeine, etc.

Nevertheless, trivial names tell us nothing about the structure. Therefore, whenever I hear some trivial name for the first time in my life, I have to ask Google how that particular molecule looks. Moreover, imagine learning all trivial names by heart, while there are millions of organic molecules to be named!

We need something more reliable, some kind of universal naming system for all possible molecules, which does not require too much memorizing.

International Union of Pure and Applied Chemistry comes with help. IUPAC is an organization, where people define rules governing communication between chemists. It includes rules of **chemical nomenclature**, what means "naming molecules."

The name created according to IUPAC rules is a so-called **IUPAC name** or **systematic name**, because there is a universal system of rules behind (there are some differences between both terms, for example, not all IUPAC names are systematic, but you do not need to bother).

Let me explain the idea. We have drawings to communicate graphically – I show you a picture and a structure of a molecule we talk about is explicit. Names are for verbal communication, and it would be sensible if their task were equal to the role of a drawing – to *describe the structure, but with words instead of sticks.* IUPAC names are just like that. They contain full information about the structure, but first, we must learn "the code", which chemists around the world use on a daily basis.

Take a look at IUPAC names of menthol and citronellol. In this chapter, our aim is to get acquainted with all these chunks, like "propan", "yl", "methyl", "en", "ol", etc.:

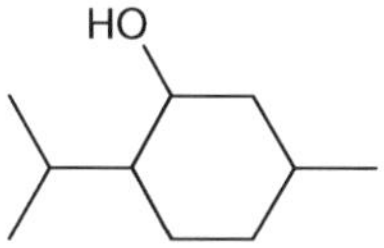

HO

HO

trivial name
menthol
IUPAC name
2-(propan-2-yl)-5-methylcyclohexan-1-ol

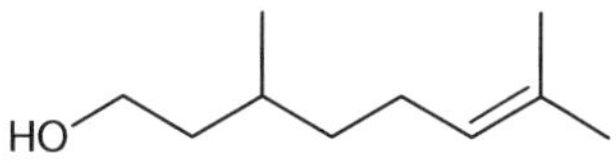

trivial name
citronellol
IUPAC name
3,7-dimethyloct-6-en-1-ol

Fig. 7.1

Every IUPAC name is glued from various “chunks” & so-called locants

IUPAC names often look quite inhumanly, because they are blends of “chunks” and **locants**. Locant (▒) always belong to the chunk, and they go in pairs: ▒-chunk.

Every chunk is a little word, which encodes a particular fragment of the molecule. Locant is a number (from 1 up), which tells us where the fragment is located. We leave locants for later. In this chapter, we learn about chunks:

1. chunks to encode functional groups
2. multiplying chunks for multiply-occurring motifs
3. chunks to encode branches of the backbone
4. chunks to encode the structure of the root backbone

Chunks to encode functional groups

In the table below, I have ordered functional groups according to their priority and added corresponding chunks. Bold font marks these, which you should learn by heart; the rest is rarely used.

How to use the table? Let us say you want a chunk to encode the presence of an OH functional group. Whenever *it is* the highest priority group in your molecule – select **ol**. Whenever an OH group *is not* the highest priority one – choose a **hydroxy** chunk. Check row 8:

Pr.	Functional Group		Chunk if not the highest priority	Chunk when FG wins in priority
1st	COOH	carboxylic	-	**oic acid**
2nd	COCl	acid chloride	chlorocarbonyl	oyl chloride
3rd	COO*C'*	ester	*C'*oxycarbonyl	***C'*yl oate**
4th	$CONH_2$	primary amide	carbamoyl	**amide**
5th	CN	nitrile	cyano	nitrile
6th	CHO	aldehyde	formyl	**al**
7th	C=O	carbonyl	oxo	**one**
8th	OH	hydroxyl	**hydroxy**	**ol**
9th	NH_2	primary amine	amino	**amine**
10th	=NH	primary imine	imino	imine
11th	O	ether	oxy	ether
12th	C=C	alkenyl	**en**	
13th	$C\equiv C$	alkynyl	**yn**	
14th	aromatic benzene ring		**phenyl**	**benzene**
15th	X	halogen	**fluoro, chloro, bromo, iodo**	-

Multiplying chunks

You already know them. We use multiplying chunks *for functional groups and branches, which occur more than once in a molecule*:

2	3	4	5	6	7	8	9	10
di	**tri**	**tetra**	**penta**	**hexa**	**hepta**	**octa**	**nona**	**deca**

For instance, when a molecule has three carbonyl groups as the highest priority ones, a multiplying chunk **tri** will precede the chunk **one**, which encodes C=O group. The result will be **trione**, which means "three carbonyl groups."

Problems

Write chunks of functional groups, which should appear in IUPAC names of the following molecules:

7.1

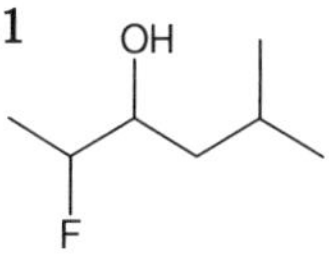

7.2

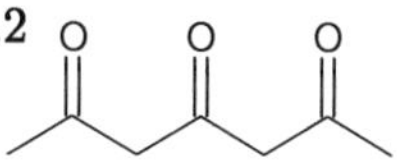

7.3

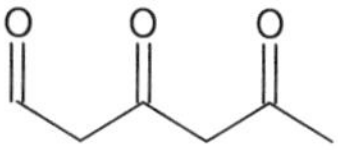

7.4

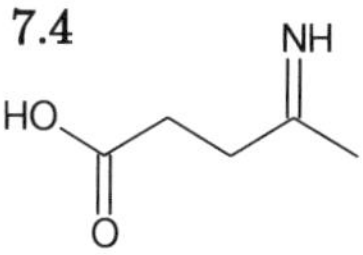

7.5

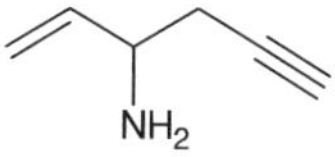

7.6

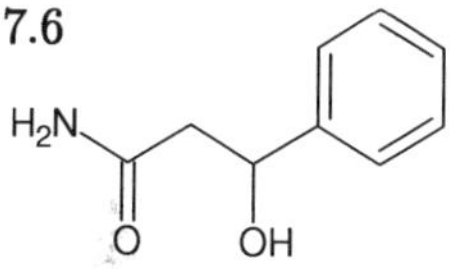

7.7

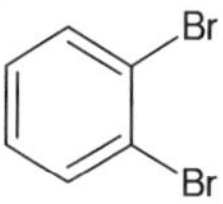

7.8

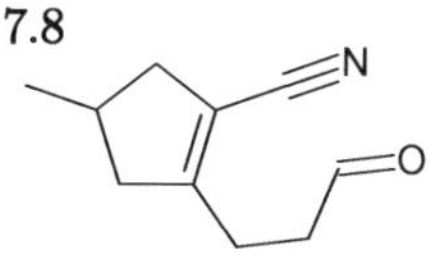

7.9

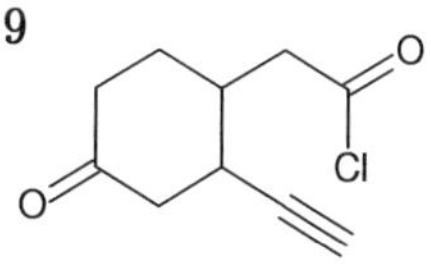

Clusters of chunks to encode structures of branches

BEGINNING...	...MIDDLE...	...END OF THE CLUSTER
cyclo chunk (present or absent)	one of the **meth...dec** series of chunks	**yl** or **an-▒-yl**

A **cyclo** chunk appears at the beginning of the cluster, when the branch is cyclic. **Meth** to **dec** are carbon count chunks. They tell us the total number of C atoms in the branch:

1 C	2 C	3 C	4 C	5 C	6 C	7 C	8 C	9 C	10 C
meth	**eth**	**prop**	**but**	**pent**	**hex**	**hept**	**oct**	**non**	**dec**

The cluster for a branch always ends with **yl** chunk. It will be the case whenever the branch is attached to the root backbone *via its 1st C atom*. However, sometimes the cluster ends with **an-▒-yl**, which means the branch bonds to the root backbone via other than its 1st C atom. For example, the cluster of a branch attached to the root via its 2nd C atom will end with **an-2-yl**.

Imagine you look at some IUPAC name, and you spot a **propyl**. This part of the name must describe a branch, because it ends with **yl**. Moreover, the branch bonds to some root backbone via its 1st C atom; otherwise, there would be **an-▒-yl** at the end. Also, it must be a chain (there is no **cyclo** chunk at the beginning), and it must be made of 3 C atoms (because the carbon count chunk is **prop**):

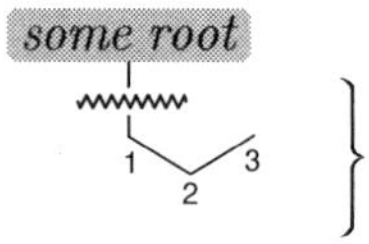

some root

1 2 3

propyl *branch*

wave over a bond always means we do not know what is at the bond's second end

Fig. 7.2

Another example is **pentan-2-yl**, which we can divide into chunks: **pent an-2-yl**. This is a 5 C chain branch (**pent**), but this time attached to some root backbone via its 2nd C atom, as the ending implies:

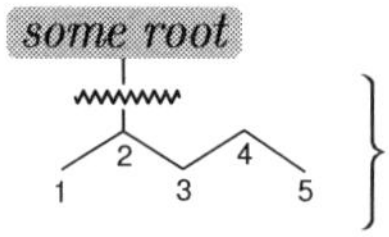

some root

1 2 3 4 5

pentan-2-yl *branch*

Fig. 7.3

And what if you spot a **cyclopentyl** somewhere else? This branch is attached to the root backbone via its 1st C atom; otherwise, there would be some **an-▒-yl** at the end. It must be a ring (**cyclo**) made of 5 C atoms (**pent**). Nobody said branches could not be cyclic:

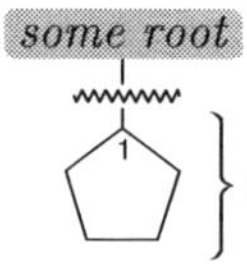

some root

1

cyclopentyl *branch*

Fig. 7.4

Common sense tells us there is no 1st C atom in the ring, because rings do not have a beginning or end. However, IUPAC rules describe in an arbitrary manner, the C atom of the ring which bonds to the root, as 1st.

Clusters of chunks to encode the structure of the root backbone

BEGINNING...	...MIDDLE...	...END OF THE ROOT'S CLUSTER.
cyclo chunk (present or absent)	one of the **meth...dec**	**an**, ▒-**en**, ▒-**yn**, or ▒-**en**-▒-**yn**

The rules are similar to those for branches. The root's cluster contains a **cyclo** chunk only when the root is cyclic, while **meth...dec** encodes the total number of Cs. As you see, the only difference between clusters for branches and root's cluster is the ending.

Root's cluster ends with **an**, ▒-**en**, ▒-**yn** or ▒-**en**-▒-**yn**. It depends on type of carbon-carbon bonds within:

only C–C bonds in the root (e.g. alk**an**es)	at least one C=C in the root (e.g. alk**en**es)	at least one C≡C in the root (e.g. alk**yn**es)	at least one C=C *and* at least one C≡C in the root
an	▒-**en**	▒-**yn**	▒-**en**-▒-**yn**

We serve locants ▒- together with **en** and **yn** chunks to avoid ambiguity. They tell us *at which C of the root a double or triple bond starts*. Without ▒-, you would not know where it is located.

Assume in some name a cluster **cyclooctan** appears. We can divide it into chunks: **cyclo oct an**. The root backbone must be made of C–C bonds only (because of **an**), must have 8 Cs (because of **oct**), and for sure it is a ring (**cyclo**). All 8 Cs are members of the ring, because *the root is never part-chain-part-cycle*:

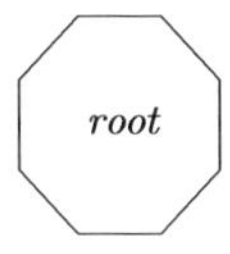

Fig. 7.5

Problems

Clusters below are fragments of real IUPAC names. Draw structures they describe. Note it can be the root or a branch. Use the wave in the latter case, to mark the bond, which joins the branch with an unknown root:

7.10 ethan
7.11 hex-1-en-4-yn
7.12 pentan
7.13 butan
7.14 cyclohexen
7.15 methyl
7.16 cyclohexyl
7.17 non-4-en-2-yn
7.18 cyclopenten
7.19 cyclopropan
7.20 pentan-3-yl
7.21 ethyl

7.22 hex-3-en	**7.23** propan-2-yl	**7.24** but-1-en
7.25 but-1-yn	**7.26** but-2-en	**7.27** heptyl
7.28 oct-4-yn	**7.29** cyclopropyl	**7.30** decyl
7.31 cyclobutyl	**7.32** methan	**7.33** ethyn
7.34 pent-2-en	**7.35** hex-1-en	**7.36** propan
7.37 heptan-4-yl	**7.38** pentyl	**7.39** but-2-yn
7.40 cyclobuten	**7.41** hex-2-yn	**7.42** hex-4-en-1-yn
7.43 pent-3-en-1-yn	**7.44** cyclopentan	**7.45** pent-1-en
7.46 octyl	**7.47** cycloheptan	**7.48** butan-2-yl
7.49 cyclopropen	**7.50** hexan-2-yl	**7.51** hexan
7.52 ethen	**7.53** butyl	**7.54** pent-2-yn

I showed you in answers, that sometimes chemists represent normal branches attached via their 1st C, with their abbreviations, instead of zigzags. 'Et' means ethyl, 'Pr' propyl, and 'Bu' is butyl. By analogy, 'Me' stands for methyl. Another useful symbol is 'Ph', which means *monosubstituted benzene ring.*

Finish names of hydrocarbons with "e" chunk

You are ready to interpret and recognize IUPAC names of *unbranched* hydrocarbons. We just need to add a special chunk "**e**" at the end of the name. In fact, you can take any root's cluster from the previous problems, add "**e**" chunk, and we end up with a correct IUPAC name of a corresponding hydrocarbon.

Problems

Draw following hydrocarbons:

7.55 hex-1-en-4-yne	**7.56** hex-3-ene	**7.57** cyclohexene
7.58 non-4-en-2-yne	**7.59** pentane	**7.60** ethene

Popular non-systematic names of branches

From time to time you will spot a name, which includes a traditional nickname of a branch. For instance, a **propan-2-yl** group is often referred to with its nickname: **isopropyl**. **Butan-2-yl** may appear as ***sec*-butyl**. In general, there are at least six branches, for which traditional nicknames are in common use:

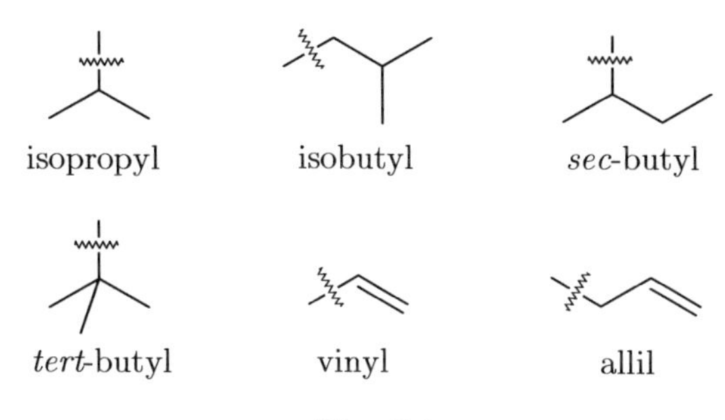

Fig. 7.6

Deducing IUPAC names for **isobutyl**, ***tert*-butyl**, **vinyl** and **allil**, may be impossible for you, because we introduce only a simplified version of organic chemistry nomenclature.

08 Translating the name into the structure

In IUPAC name all chunks & clusters
are arranged into three following sections:

-clusters for branches, -chunks for other FGs	cluster of the root backbone	-chunk for the highest priority functional group*

*use **e** chunk when root's cluster is the final part of the name

At the beginning of IUPAC name, we have all attachments listed together with their locants: all branches of the root backbone, and all functional groups, except for the one with the highest priority. We *always order this list alphabetically.*

In the middle of the name, the cluster describing the root backbone sits. It is easy to fish it out, because it must end with an, en, yn, or enyn, and start with cyclo (if cyclic) or meth...dec (when chainlike). At the end, there is chunk coding the highest priority functional group.

Examples of IUPAC names

Let us try to decode **3,7-dimethyloct-6-en-1-ol**. The task is to translate it into the correct structure drawing.

We see **1-ol**, which encodes OH, as the highest priority functional group. Then, in the middle, there is a root's cluster, which indeed ends with **en**. Where does it begin? There is no **cyclo** chunk, so we look for one of **meth...dec**, and spot **oct**. Whatever is at left from **oct** constitutes the first section of IUPAC name:

3,7-dimethyl	oct-6-en	1-ol

In **3,7-dimethyl** cluster we see two locants, and multiplying chunk **di**. It means the molecule contains two **methyls**. They are attached to 3rd and 7th carbon atom of **oct-6-en** – the 8 C chain root, with one C=C bond starting at 6th C atom. The chunk **1-ol** means that the hydroxyl group is attached to the 1st C atom of the **oct-6-en** root.

Time for drawing. We always start from the root, so the first thing to draw is the 6 C chain. Second thing you should always do is numbering root's C atoms. It helps to avoid later mistakes. Only then, we can decorate the zigzag with remaining structural moieties:

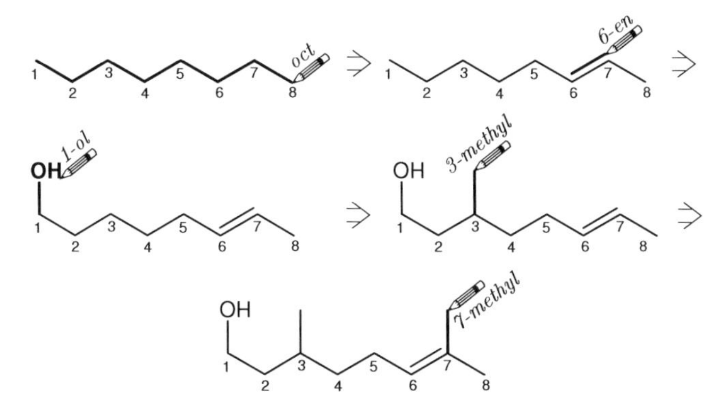

Fig. 8.1 Drawing 3,7-dimethyloct-6-en-1-ol a.k.a. citronellol.

The molecule is a familiar citronellol. Now, let's take a look at the second example. Its IUPAC name is **5-methyl-2-(propan-2-yl)cyclohexan-1-ol**. I have chopped it into sections to make things more clear:

5-methyl 2-(propan-2-yl)	cyclohexan	1-ol

The root is simple **cyclohexan**, and once again, there is a hydroxyl group attached. In the first section, we have two clusters for two different branches: a **methyl** branch at 5th C of the root and **propan-2-yl** branch, attached to the 2nd C. Note that the list is alphabetically ordered – m appears before p in the alphabet.

Interestingly, **propan-2-yl** is enclosed between parentheses. This is one of IUPAC rules – *use parentheses for any cluster, which ends with an-▒-yl, whenever there is something preceding it in the IUPAC name.*

Brackets increase clarity, and emphasize the coherence of (propan-2-yl) cluster, lest you will confuse internal "2" (which means 2nd C of the branch), with outside locants (which always refer to Cs of the root).

Now we can do the art. We start with a cyclohexan root and attach OH to the 1st C atom. All Cs are equivalent, so we choose one arbitrarily. Then we can add remaining attachments:

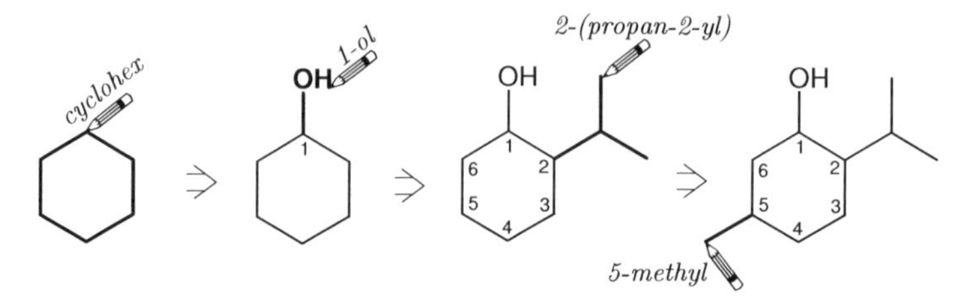

Fig. 8.2 Drawing 5-methyl-2-(propan-2-yl)cyclohexan-1-ol (menthol).

Problems

Look at IUPAC names below and draw corresponding molecules:

8.1 draw 2,3-dimethylpent-2-ene
8.2 draw 1,1,2,3-tetramethylcyclopentane
8.3 draw 1-methyl-2-pentylcyclobutane
8.4 draw 5-*tert*-butyl-3-methylcyclohex-1-ene
8.5 draw 4-hydroxypentan-2-one
8.6 draw 3,4-dimethylhexan-2-one
8.7 draw 2-chloro-2-methyl-6-propylcylohex-3-en-1-one
8.8 draw 3,4-dibromo-6-phenyloct-3-en-2-ol
8.9 draw 2-chloro-2-methyl-6-propylcylohexane-1,4-dione
8.10 draw hept-3-en-yn-2-one (note: no attachments!)
8.11 draw 2-phenyl-5-propylcyclopentan-1-imine
8.12 draw 2-amino-3,3-diphenylpropan-1-ol
8.13 draw 2-*sec*-butylcyclopent-3-en-1-amine
8.14 draw 1,2-dimethylcylohepta-4,6-diene-1,3-diol
8.15 draw 3-ethyl-4-imino-6-methyloctan-2-ol
8.16 draw 1-cyclopentylpental-2-ol
8.17 draw 7-cyclopropyl-6-ethyloct-5-en-4-one
8.18 draw 5-(butan-2-yl)-2-cyclopentylcylohex-3-en-1-amine
8.19 3-ethyl-1-fluoro-2,5-dimethyl-4-propylcyclohexa-2,5-dien-1-amine

8.20 Note that all attachments in the first section are listed in alphabetical order, but multiplying prefixes *never count.* That is why in molecule **8.19** we have: **ethyl** followed by **fluoro**, while **dimethyl** appears as last. Also, recall that we can draw molecules with abbreviations of branches instead of zigzags (e.g. Me, Et, Pr, Bu...). Redraw molecule **8.19** using abbreviations for branches, instead of zigzags.

09 Naming the molecule step by step

Now it is time for you to try the reversed approach – to translate the structure drawing into IUPAC name. Fortunately, you already know most stuff needed. However, two new skills are required: distinguishing the root from branches, and numbering Cs in the root. And IUPAC rules are strict – for each molecule there is only one correct root, and only one correct way to number Cs inside. Other choices are mistakes, which ruin the name.

Before we start, read a description of steps we take, when we generate IUPAC name. Do not expect to understand all details now, but return to this recipe when necessary:

A general recipe to name a molecule step by step

A like an **a**nchor – find the most important functional group

B like **b**ackbone – distinguish the root from branches

C like **c**arbons – number Cs in the root

These first three stages prepare us to name a molecule. They are crucial – most incorrect IUPAC names students happen to concoct, are direct results of errors made at the stage B or C. Subsequent steps are less demanding:

1 Organize all chunks and clusters, together with their locants, into three sections: 1) attachments, 2) root's cluster, and 3) the highest priority group. At this step:
- remember to join multiply-occurring motifs into one chunk, with the help of multiplying chunks,
- order the list of attachments alphabetically,
- add chunk "e" if there is no "third section."

2 Glue all that into one-word IUPAC name, using hyphens "-" to separate any number from any letter.

3 Whenever applicable, add stereodescriptors* at the mere beginning (*stereodescriptors will be introduced in Part IV).

To correctly distinguish the root from branches is the first step to success

The root will always be a chain or a ring. Look at rules to follow:

Primary rule Root backbone must have the highest priority functional group attached.

Secondary rule Root backbone should contain maximum number of C=C and $C\equiv C$ bonds.

Tertiary rule The root should be as large as possible.

Primary rule is very useful at the start. In the molecule below an amine group, NH_2, is the highest priority functional group. It must be attached to the root, so first four Cs are obvious. Then we meet a branching point. Where the root continues? Which of two chains is the branch? Secondary rule applies here. The root should contain C=C:

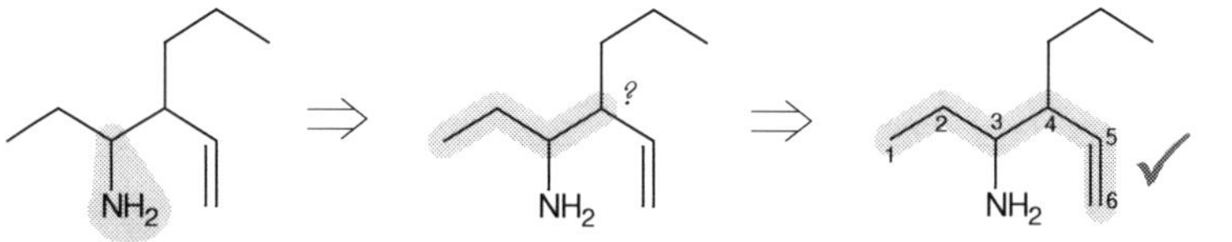

Fig. 9.1 Finding the root backbone according to IUPAC rules.

Having the root found, we can number Cs, and then write the root's cluster. It is **hex-5-en**. Next, we can encode the presence of the NH_2 group with a **3-amine** chunk, while the branch goes as **4-propyl**:

4-propyl	hex-5-en	3-amine

The final stage is to connect all chunks into one-word IUPAC name. Always use a hyphen "-" to separate any digit from any letter that appears. The result is **4-propylhex-5-en-3-amine**.

Consider the next example. The only difference is that there is no double bond, so secondary rule does not apply. However, tertiary rule says, that the root backbone should be as large as possible. Therefore, we choose the longer version:

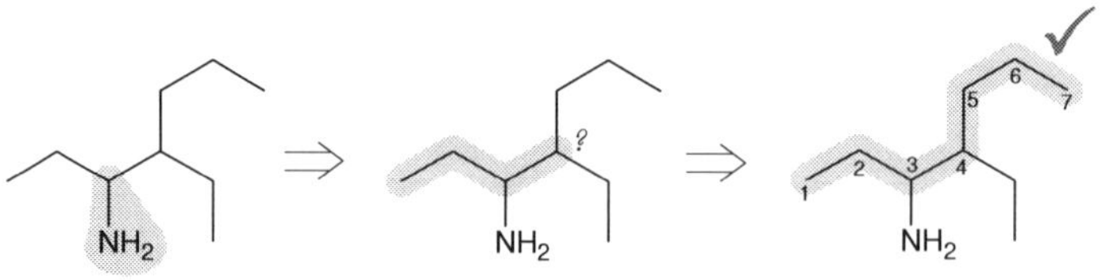

Fig. 9.2

4-ethyl	heptan	3-amine

IUPAC name of this amine is **4-ethylheptan-3-amine**.

What if there are no N or O-containing functional groups to guide us at the mere beginning? This happens in all hydrocarbons. In such situations, we just *look for the longest chain.* You should be careful, because sometimes it is a bit tricky. For example, I have drawn an alkane below in such a way, that horizontal chain simulates being a root. It is just an illusion. You had better count atoms – in numbers we trust:

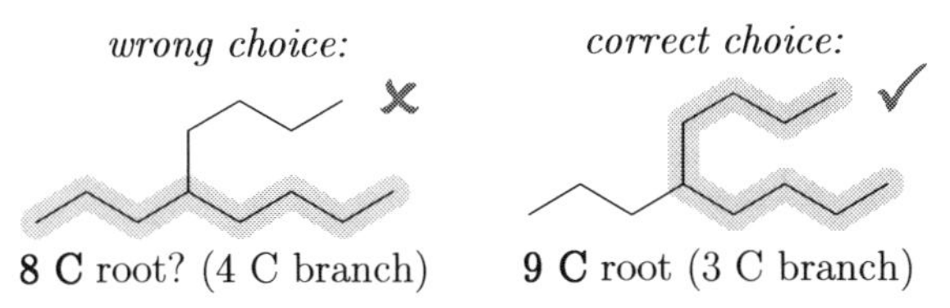

Fig. 9.3 IUPAC name of this hydrocarbon is **5-propylnonane**.

Let us now consider a bit more complex molecule. Below, we have an OH group, which wins in priority for sure. It points out the root, which must contain a C=C bond:

Fig. 9.4

Now, having the highest priority FG and the root correctly found, we number C atoms in the root from 1st to 8th. Be careful now! The only correct way to do it, is from right to the left, in order that an OH group gets locant 2. Numbering Cs from left to right would give it locant 7, and that would be a mistake. *We always check both ways and choose the one, which grants the highest priority group the lowest locant.*

Having numbers in place, we can code the root structure as **oct-6-en** and OH as **2-ol**. The only thing we still need is the list of all other attachments: a bromine atom (chunk: **5-bromo**), a **6-propyl** branch, and two **methyl** groups at 4th C.

The latter must appear in the name as **4,4-dimethyl**. I have used multiplying chunk **di**, but what is more important to notice is "**4,4**". Remember, that we *always list and separate locants with commas, even if they are equal.* The cluster **4-dimethyl** would be incorrect.

As you remember, we order attachments alphabetically, but multiplying chunks do not count:

5-bromo, 4,4-dimethyl, 6-propyl	oct-6-en	2-ol

We end up with **5-bromo-4,4-dimethyl-6-propyloct-6-en-2-ol**.

Problems

Generate IUPAC names of following molecules and determine the class of organic compound to which they belong:

9.1 9.2 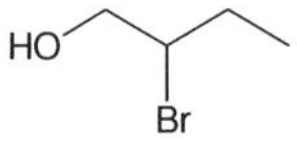 9.3 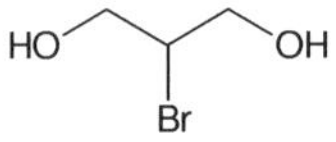 9.4

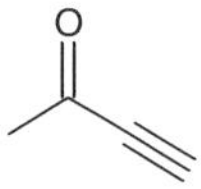

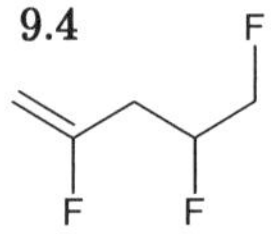

9.5 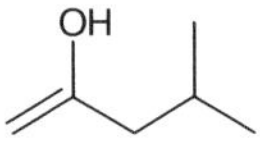 9.6 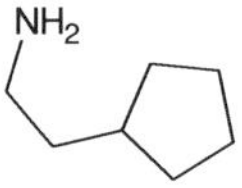 9.7 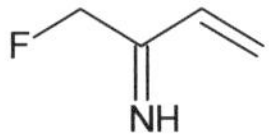 9.8

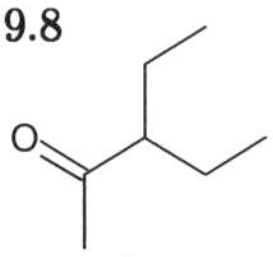

9.9 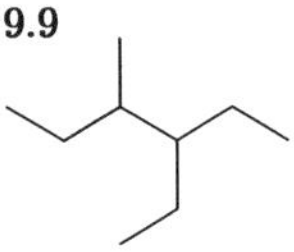 9.10 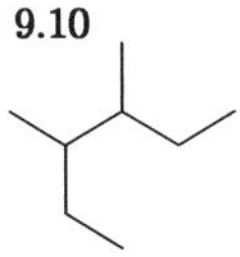 9.11 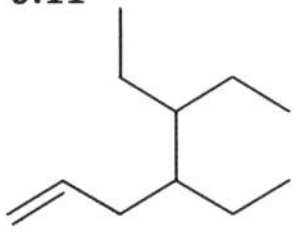 9.12

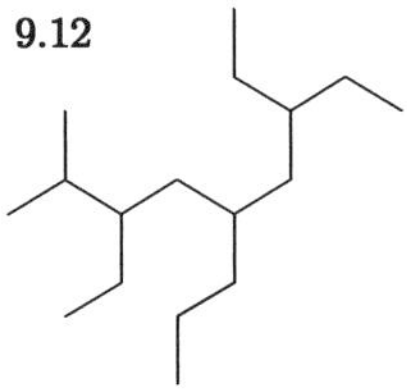

Alkane **9.12** requires more caution. Remember that the root backbone should be the longest chain, but also *branches must be as simple as possible*, so that we can easily name them.

9.13-9.17 There are 5 isomeric alkanes with C_6H_{14} formula. Draw all of them and write their IUPAC names.

To correctly number C atoms in the root is the second step to success

There is no such a claim that we should number Cs from the left side to the right side. Some students tend to think so, because this is how they read and write sentences. Anyway, it would be ridiculous, because in such situation rotating or flipping the drawing would change values of locants in the name of a molecule.

As I said before, for each chainlike root, there are two directions of numbering possible – from one end to the other or vice versa. Your job is to check both ways, and choose the one, which grants the highest priority FG the lower locant. In other words, start numbering from the end of the chain, which is closer to the highest priority group.

Problems

Determine IUPAC names of following molecules:

9.18 9.19 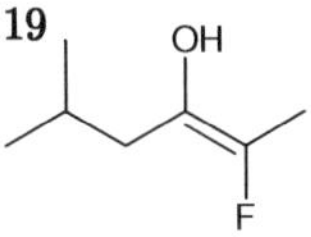 9.20

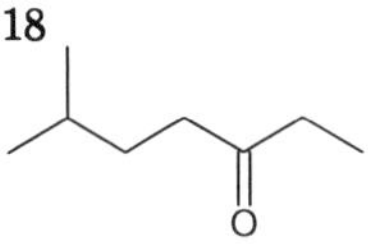

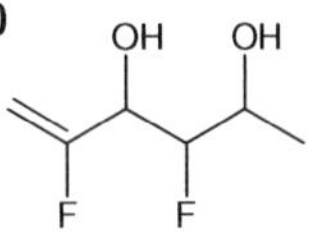

9.21 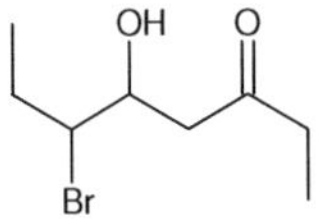 9.22 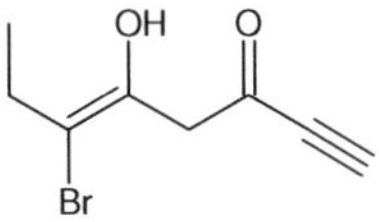 9.23

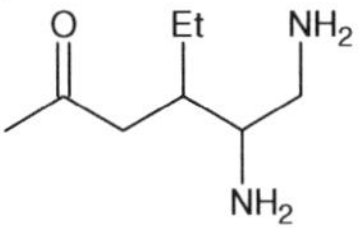

9.24 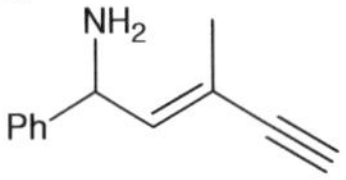 9.25

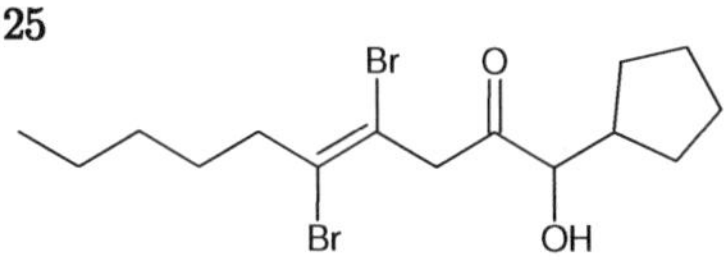

What if the highest priority group sits in the middle of the chain, and therefore gets the same locant irrespectively to how you number Cs? In such cases we look at all other attachments, and choose the direction of numbering, which gives *them* lower locants.

Problems

Determine IUPAC names of following molecules:

9.26 9.27 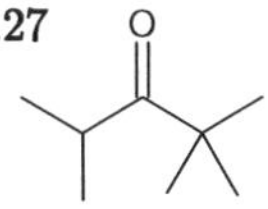 9.28

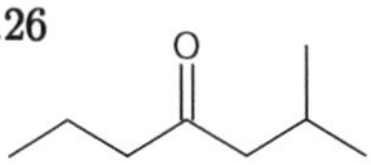

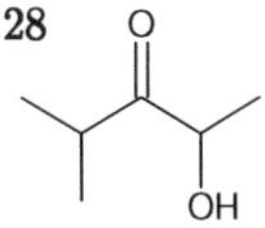

In case of cyclic roots, there is a bit more hassle with numbering Cs, because rings have no beginning to start from.

The rules say that we should assign the locant "1" to that C atom of the ring, which bonds to the highest priority group (recall that, on the other hand, when the cycle is a branch, we assign locant "1" to C atom, which bonds to the root).

Numbering goes clockwise or anticlockwise and we choose the option, which gives lower locants to remaining substituents:

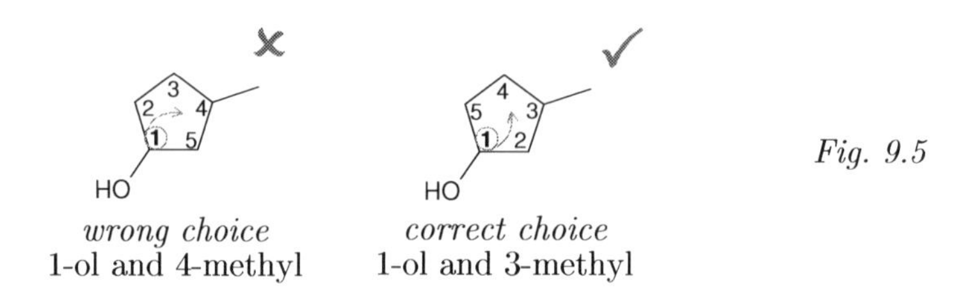

wrong choice
1-ol and 4-methyl

correct choice
1-ol and 3-methyl

Fig. 9.5

Therefore, chunks for the above alcohol look like this:

3-methyl	cyclopentan	1-ol

After gluing it, we get IUPAC name: **3-methylcyclopentan-1-ol**.

A similar molecule, but without the methyl group, would be referred to as cyclopentanol, instead of cyclopentan-1-ol. *We can omit "1-" in names of molecules with monosubstituted rings*, because it causes no ambiguities.

Problems

Determine IUPAC names of the following molecules:

9.29 NH_2 **9.30** OH NH_2 **9.31** NH_2 **9.32** OH

9.33 OH **9.34** O **9.35** O

9.36 O Et Me **9.37** NH **9.38** NH_2 **9.39**

9.40 O O **9.41** Bu **9.42** OH Ph **9.43**

9.44-9.55 There are as many as 12 structurally isomeric cycloalkanes with C_6H_{12} formula. Draw all of them and write their IUPAC names.

Name ketones below. Popular non-systematic nicknames of branches (Chapter 7) *may be necessary* (or maybe not?):

9.56 9.57

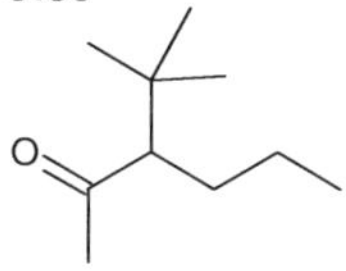

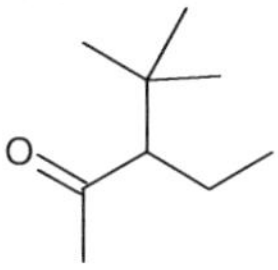

There are functional groups, which appear with no locant, whenever they win in priority:

COOH, COCl, COO*C'*, $CONH_2$, C≡N, CHO

Whenever one of these groups is the highest priority functional group, *we treat its C atom as part of the chainlike root*, and assign a locant "1" to it. Later, we *omit* the locant in the name.

Problems

Draw following molecules:

9.58 4-aminopentanoic acid
9.59 2,3-dimethyl-4-phenylpentanal
9.60 4-amino-3,3,5,5-tetramethylhexanoic acid

Determine IUPAC names of following molecules:

9.61 O OH

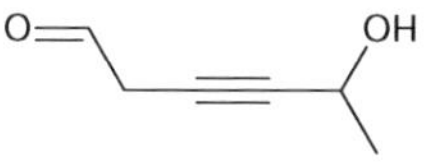

9.62 N O

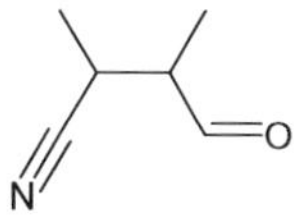

9.63 O Cl F

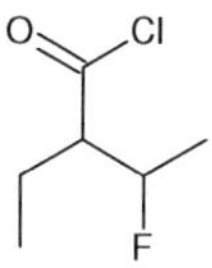

9.64 O NH_2 F

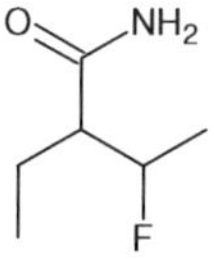

9.65 N

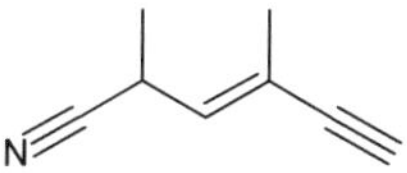

9.66 O

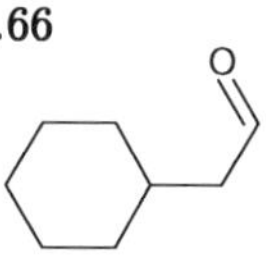

Naming esters requires separate commentary

Among carbonyl compounds, we have an ester functional group, COO*C'*. Esters name is two words, and ends with a characteristic chunk **oate**. Consider following examples, and try to understand how we generate such names:

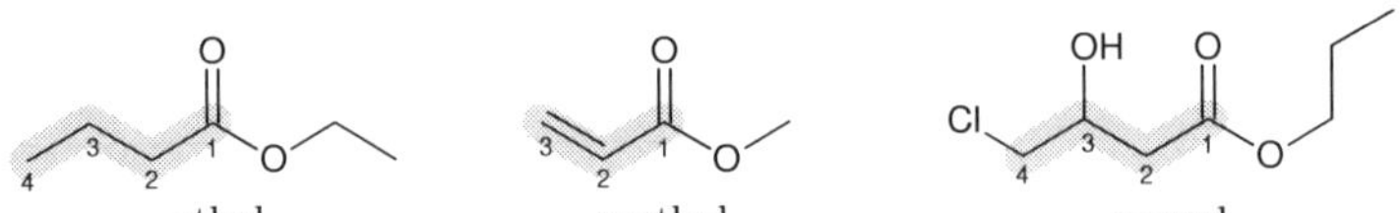

ethyl
butanoate

methyl
prop-2-enoate

propyl
4-chloro-3-hydroxybutanoate

Fig. 9.6

"Right side" of COO*C'* is usually very simple, and encoded by the first word of the name. We name it as if it were a branch on the root backbone (methyl, ethyl, propyl, etc.).

Second word is a standard IUPAC name generated for "the left side" of the ester. We treat left C in COO*C'* as the first atom in a chainlike root. In the name, we omit its imposed locant "1".

Problems

Draw the following esters:

9.67 methyl 3-aminobutanoate
9.68 butyl 5-hydroxy-2,3,4-trimethylhexanoate
9.69 propyl 2-cyclopropylpropanoate

Determine IUPAC names of the following esters:

9.70

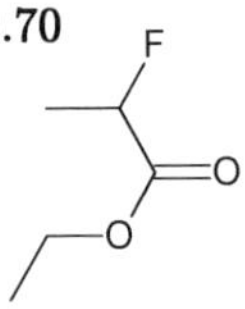

9.71

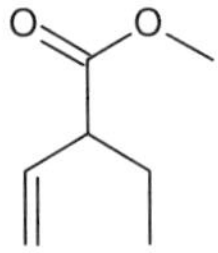

9.72

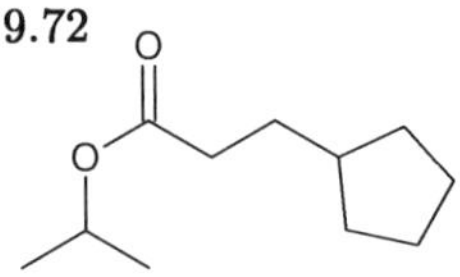

Common names of organic molecules

Some organic compounds are known for so long, that their common names are older than the entire IUPAC nomenclature. They are still widely used, and more popular than corresponding systematic names.

IUPAC treats few of them as "preferred non-systematic IUPAC names." Nevertheless, these are nuances. What is more important, is that you will need to remember some of these. How much? It depends on your teacher's requirements:

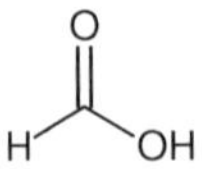

formic acid
(methanoic acid)

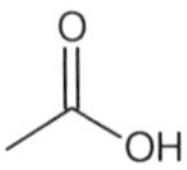

acetic acid
(ethanoic acid)

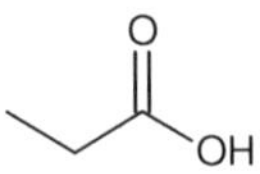

propionic acid
(propanoic acid)

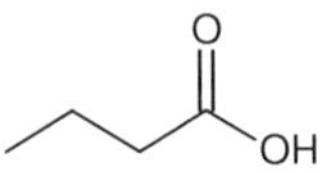

butyric acid
(butanoic acid)

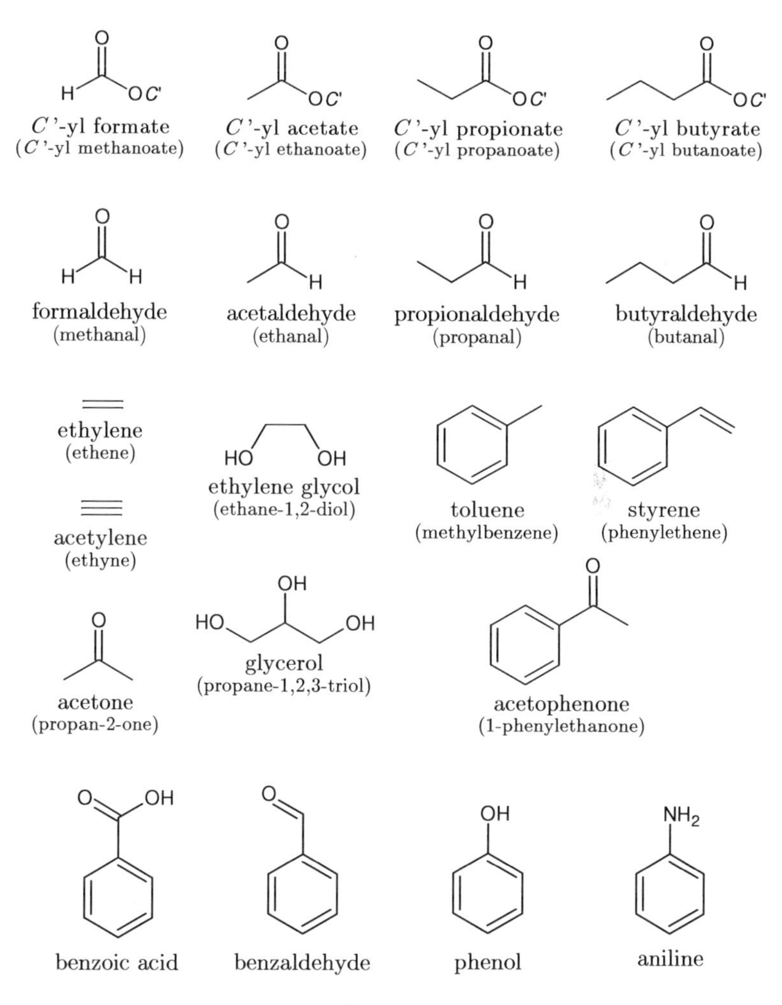

Fig. 9.7

When IUPAC nomenclature becomes cumbersome

Algorithmic IUPAC procedure for naming molecules works well for small and medium sized molecules. However, no one who is of sound mind tries to name large ones. Since a systematic name contains all structural information, its size and intricacy grow with the size of a molecule. At some point, it becomes so cumbersome, that we must forget about systematic nomenclature.

In biological and medical sciences, we often deal with large molecules and no one uses long, illegible IUPAC names. Imagine a biochemistry textbook if we were forced to do so! Many common names exist and will never die out, simply because they are handy.

Part III CONFORMATIONS

10 How to deduce the 3D shape of the molecule

In last chapters, we drew molecules as 2D objects, though we know that they are 3-dimensional in fact. The branch of organic chemistry, which focuses on 3-dimensionality of molecules is **stereochemistry**.

We start from deducing the 3D shape of a molecule, and later delve into other stereochemical considerations in this and the next part of the book.

Pairs of electrons repel each other

Imagine you have a chunk of modeling clay. Tear some, form a ball, and pretend it is a C atom. Take four matches. We are going to make a model of a methane molecule, CH_4. Matchsticks represent chemical bonds, while their heads are H atoms. Make a 3D model of methane by pressing each of four matches into the clay ball.

Do you know the 3D shape of CH_4? Alas. The problem is that you can arrange four matches around a ball in a theoretically infinite number of ways. Each of the "shapes" has different distances and angles between C–H bonds. Which one of them is the best representation of the reality?

We can deduce it, because *all pairs of electrons repel each other.* For the sake of simplicity, let us agree that it is so because all electrons carry an electric charge of -1, and there is always electrostatic repulsion between charges of the same sign. In consequence of this repulsion, distances among all pairs of electrons are maximized in real molecules.

Electron pairs sprouting in 4 directions adopt tetrahedral geometry

Let us assume our clay model of methane is planar (flat), like model **I** in the figure below. Distances are not yet maximized – simply our model does not make use of all 3D space around, and there is some room below and above. We had better move a bond out of plane (model **II**). That immediately creates an opportunity to increase the distance between remaining three. Arrange them triangularly to obtain model **III**. Now, we are in a better situation than at the beginning, though there is still one step to perfection. Send three in-plane bonds down. It moves them even further away from the vertical bond, and from each other too:

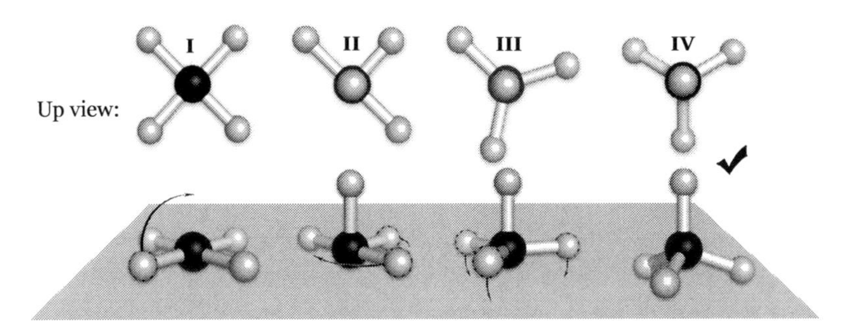

Fig. 10.1 Adoption of tetrahedral geometry.

The work is done. Model **IV** represents so-called **tetrahedral geometry** around the C sp^3 atom, and it is the real 3D shape of CH_4. Distances between electron pairs are maximized.

Look at ethane, CH_3CH_3, rotating before your eyes. Here too, all chemical bonds around both C sp^3 atoms arrange tetrahedrally. It will always be like that. Tetrahedral geometry is a common feature of all atoms sprouting electron pairs in 4 directions:

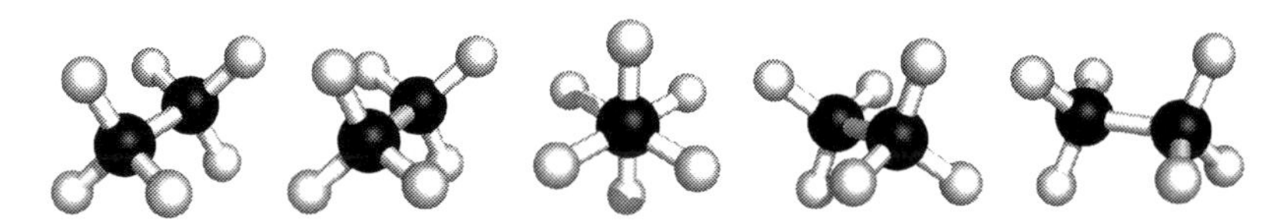

Fig. 10.2 Ethane molecule (CH_3CH_3) in 3D.

Lone pairs of electrons also count, because they also take part in mutual repulsion and need some space to occupy. Therefore, tetrahedral arrangement of electron pairs is in fact characteristic of all sp^3 hybrids:

C sp^3 (4 single bonds),
N sp^3 (3 single bonds + 1 lone pair),
O sp^3 (2 single bonds + 2 lone pairs),
and halogen atoms (1 bond + 3 lone pairs).

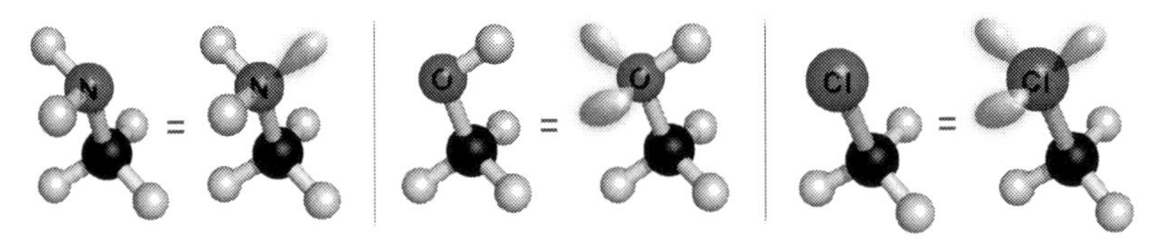

Fig. 10.3 Methanamine, methanol and chloromethane molecules. Teardrops represent space where electrons of lone pairs flit about.

On professional structure drawings, we draw carbon chains, even if they have N or O atoms inserted, as zigzags. This is because tetrahedral arrangement of bonds makes longer chains zigzag-shaped indeed.

Electron pairs sprouting in 3 directions adopt triangular geometry

When an atom has a double bond (sp^2 hybrid), things become different. Take a look at ethene, $CH_2{=}CH_2$. Both C atoms have electron pairs sprouting in 3 directions. In such situations, the best shape to adopt is a flat, **triangular geometry**:

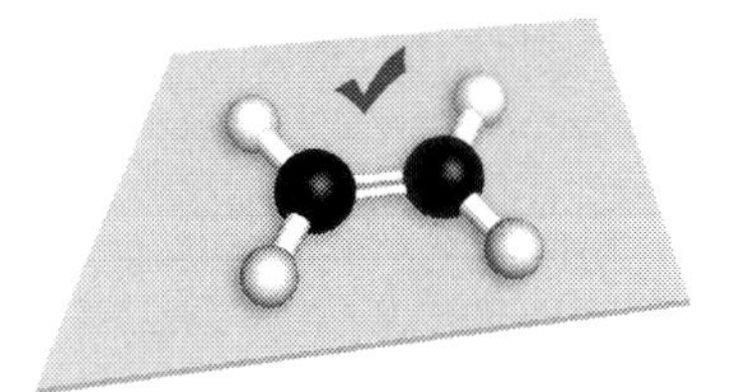

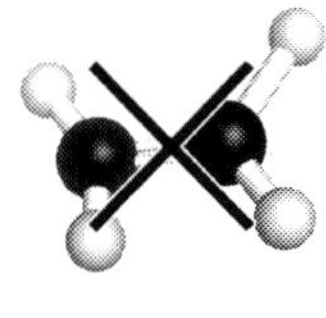

Fig. 10.4

Ethene is planar, and geometry around C atoms is triangular.

It concerns not only C sp^2, but also N sp^2 (1 single bond + 1 double bond + 1 lone pair) and O sp^2 atoms (1 double bond + 2 lone pairs).

Electron pairs sprouting in 2 directions adopt linear geometry

In ethyne $CH{\equiv}CH$, each C sp atom sprouts one triple and one single bond. They repel each other, and only **linear geometry** guarantees maximization of distances. Therefore, every molecule is straight around the part with $C{\equiv}C$ bond, which explains why we draw it as such on professional structure drawings:

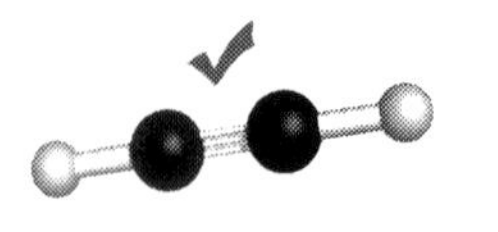

Fig. 10.5

Ethyne is linear. Any bend puts bonds closer to each other.

Linear geometry is characteristic of C sp, as well as N sp atoms.

To deduce overall 3D shape of a molecule we must consider the geometry around every atom

Any time you try to figure out the 3D shape of a molecule, just consider each of its atoms. In the figure below, I drew 3D models of three molecules – propan-2-one, also known as **acetone**, ethanenitrile better known as **acetonitrile**, and ethyl ethanoate, also known as **ethyl acetate**. These three compounds are popular organic solvents, and because of a long-lasting history, their common names remain in use.

Consider these examples to convince yourself, that imagining 3D shape of a molecule is not at all difficult. We just need to localize atoms with tetrahedral geometry (I have encircled them); flat triangular geometry (mind gray triangles), and determine which atoms adopt linear arrangement (a dashed rectangle):

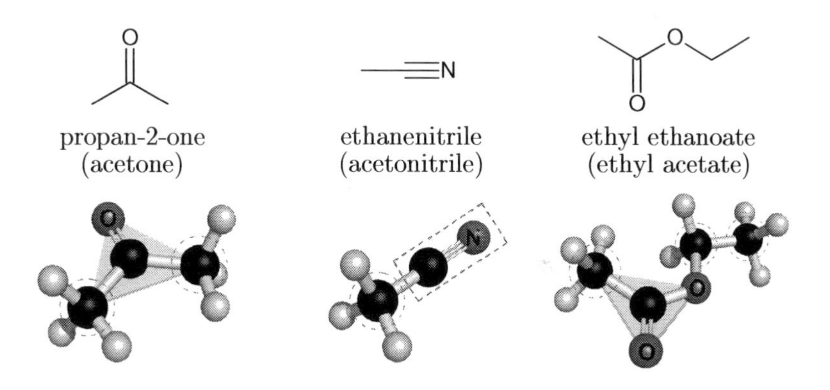

Fig. 10.6 3D shapes of the simplest ketone, nitrile and one of esters.

We can even attempt to draw structures resembling 3D features, though it becomes hard, when the complexity of structure increases:

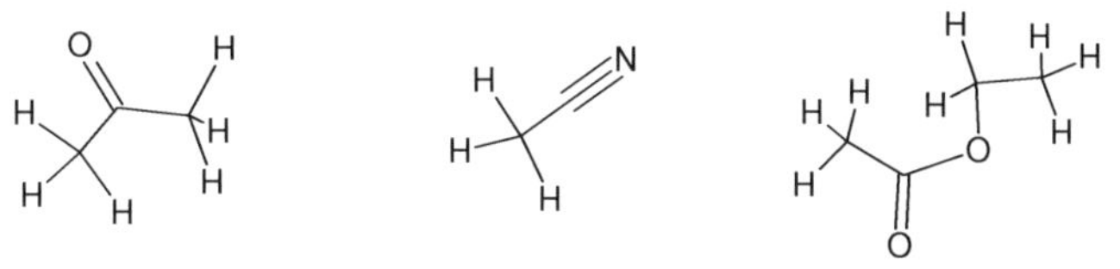

Fig. 10.7 Drawings with 3D features.

Problems

In the following molecules, mark all atoms according to the geometry adopted by electron pairs around them. Follow the example:

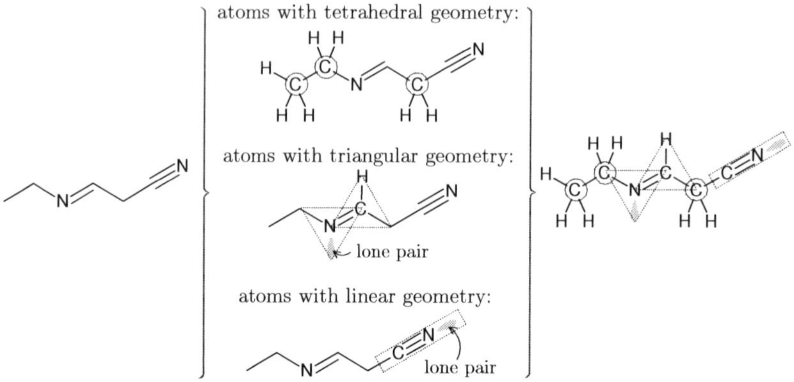

atoms with tetrahedral geometry:

atoms with triangular geometry:

lone pair

atoms with linear geometry:

lone pair

10.1

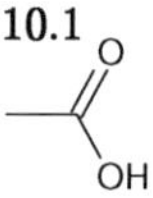

10.2 $CHCl_3$

10.3

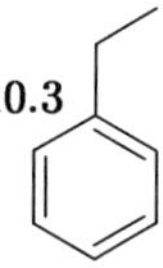

10.4

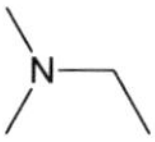

10.5

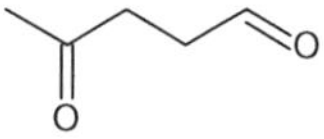

10.6

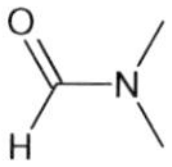

10.7

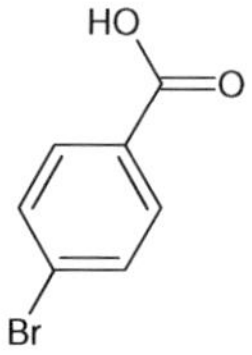

10.8

10.9

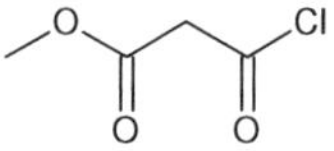

10.10

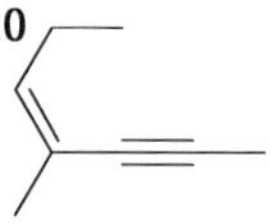

11 Conformations and Newman projections

Molecules are dynamic objects – not stone sculptures. In fact, *molecular drawing is like a snapshot cut out from an animation* showing relentless internal movements. Like a photograph, it stays motionless all the time, while in reality things are different. Unfortunately, we do not have animated pictures in books to show what our inherently dynamic molecules do all the time, so I will need to describe it with words.

Dynamism within a molecule is a result of rotations around single bonds

Imagine a model of ethane, CH_3CH_3, made of two clay balls joined with a match. Each ball has three other matches stuck in, to simulate C–H. All that arranged in a tetrahedral way:

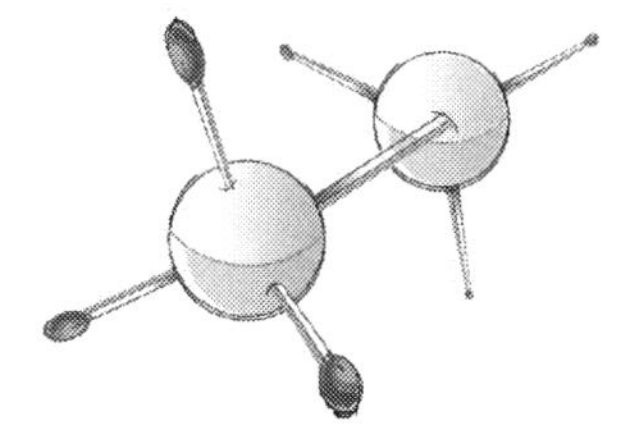

Fig. 11.1

Homemade model of ethane molecule.

Can you rotate clay balls along the central matchstick? Sure! Imagine you grab one clay ball with your left-hand fingers and another with your right-hand fingers. Now turn clay balls around C–C bond. It works and does not hurt your molecular model. Atoms are still bonded in the same way, and tetrahedral geometry lasts.

Rotation *changes neither the structure of the molecule nor the favorable geometries around atoms.* The only things, which change, are relative positions and distances between substituents at the C–C bond, in this case six H atoms. Please, note that in case of double and triple bonds, *rotations are not possible* (imagine you hold a clay-matchstick model of, let's say, ethene $CH_2{=}CH_2$; try to twist it; the balls would rather smash).

The number of possible relative positions of CH_3 groups in ethane is virtually infinite. In the figure below, I have drawn two of them. Note that the right picture is a result of a 60° clockwise rotation of the back CH_3 group along the C–C bond axis:

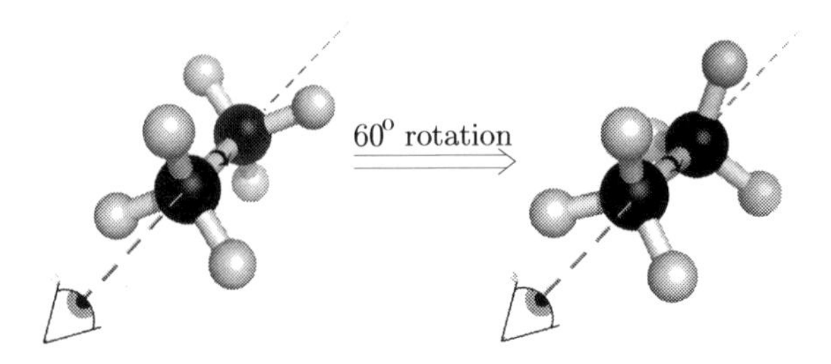

Fig. 11.2 Constant rotation around single bonds changes relative positions & distances between atoms in 3D space.

These two "shapes" are special, because they stand out from a virtually infinite set of possibilities: the left has maximal distances between C–H bonds, while in the right one these distances are minimal. Imagine you rotate any of CH_3 groups continuously – clockwise or anticlockwise, whatsoever – to convince yourself. Of course, formally, C–H single bonds rotate too, but it does not change anything.

Rotations around all single bonds in a molecule are the source of its inherent shape dynamism. We call all these different overall shapes of a molecule its **conformations**. Therefore, two models above depict two of many conformations of an ethane molecule. CH_3 groups revolve around the C–C bond constantly, so both pictures are just snapshots cut out from the animation of a "lively" ethane molecule.

Recall three molecules from the last chapter and, as an exercise, try to imagine their various conformations. Note that in ethyl ethanoate, changes of the overall shape are quite expressive:

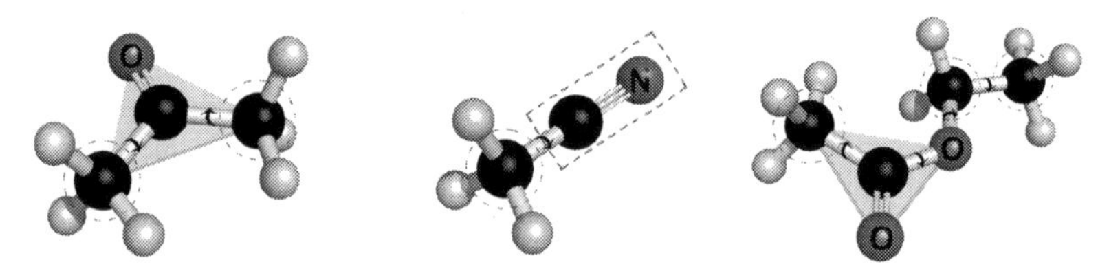

Fig. 11.3 3D shapes of the simplest ketone, nitrile and one of esters.

And what should we expect when some long chain experiences conformational freedom? All bonds can rotate simultaneously, so the overall shape changes dramatically. All of a chain looks like it is in constant contortion, and the number of various conformations is beyond our cognition.

Always think of conformations as different but interchangeable shapes of a molecule

My body has a particular, well-defined structure, and I assume yours is not different in this regard. The structure is one thing. The other is our ability to adopt various shapes.

Walking, swimming, squatting, waving your hands, snapping fingers... All these actions involve various transient shapes of our bodies, our own conformations. We have millions of them, and yogis are even better. Always think of conformations of molecules as *transient, interchangeable shapes.* Just like in the case of your body.

These various conformations of a molecule differ in stability. Our bodies are similar: standing up straight is relatively easy, assuming you are sober, while standing on tiptoes is not.

Moreover, various conformations differ in abilities. There are things, which are possible to happen only when the molecule adopts one particular conformation. It concerns many chemical transformations, but also lots of processes taking place in living organisms (for instance, when two interacting biomolecules are required to match up their shapes, so that they can fit each other like key fits the lock).

Chemists use drawings called Newman projections to show and discuss different conformations

Recall a figure of an ethane molecule as seen from several directions:

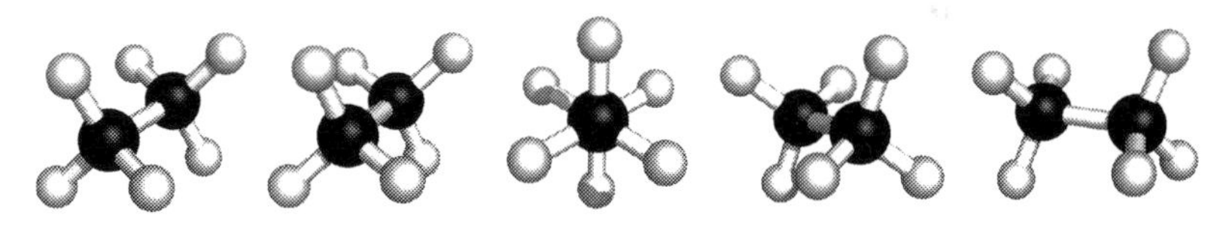

Fig. 11.4

The middle picture is a so-called **Newman projection** – *a drawing of a molecule, made from such a direction, that one of the bonds is in the axis of your gaze.* Here, it is the C–C bond. We do not see it, neither the back C atom, because the frontal one obscures them. Newman projection is the most comfortable "point of view" to observe rotation of substituents at both ends of the bond.

Let's see how it works. We are going to translate two special conformations of ethane from Fig. 11.2, into hand drawn Newman projections. In order to draw Newman projection, look at the left conformation so that C–C bond is in the axis of your gaze. This is exactly what the symbol of an eye shows:

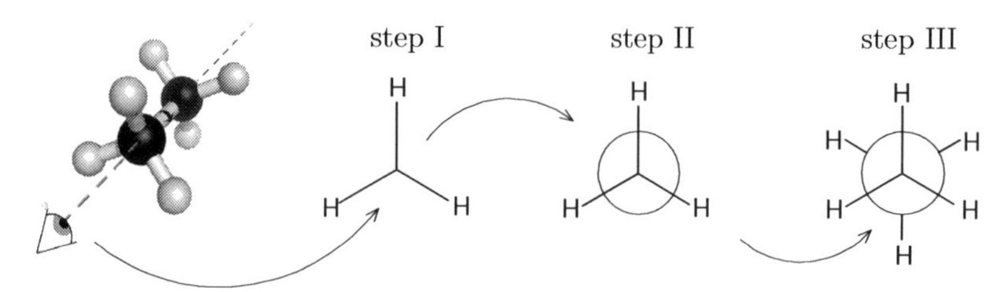

Fig. 11.5 How to draw Newman projection of one of conformations of CH_3CH_3.

We see the first C atom, and draw it as a point, followed by its substituents, three H atoms. Then, we draw the circle symbolizing the fact that first C atom obscures what is behind (step **II**), and three back substituents, which remain visible nevertheless.

Having the first Newman projection drawn, you can easily draw another, by using your imagination to push back substituents clockwise or anticlockwise. It rotates the back CH_3 group along the C–C bond axis. Note that C–H bonds in both CH_3 groups are closer and closer during the process, until you reach a second special conformation – the one with minimized distances between C–H bonds:

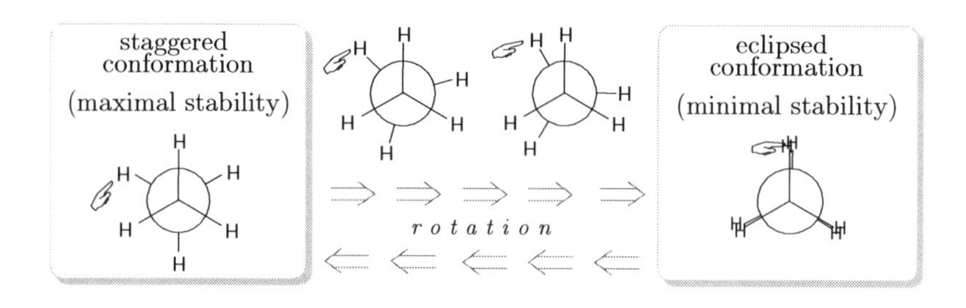

Fig. 11.6 Both models of conformations of ethane from Fig. 11.2, translated into Newman projections.

Both conformations of ethane are special: we call left one **staggered conformation**, and it is the most stable one. Right one is **eclipsed conformation**, and its hallmark is highest instability (minimal stability).

Between both special conformations, there are a virtually infinite number of transitory conformations. Stability slowly decreases when we go from staggered to eclipsed extremities. Naturally, after the molecule finds eclipsed conformation, further rotation does not provide any new insights. Stability slowly increases, as we regenerate "shapes" already seen, and after a while, we end up with a new drawing of the staggered conformation.

Why do different conformations differ in stability?

We know that electron pairs repel each other. Staggered conformation is the most stable, because there is the largest distance between C–H bonds, and in consequence, repulsion between them is the smallest. Eclipsed conformation is another extreme. C–H bonds are in the closest positions, and experience strongest repulsion. Therefore, this arrangement is the least stable. In reality, each ethane molecule spends more time in the most stable staggered conformation, and just flips over the least stable eclipsed conformation. Go and see an animated gif, which clearly shows the process:

http://en.wikipedia.org/wiki/file:ethane_conformation.gif.

Molecules with more than two special conformations

Now, we will draw conformations of a pentane molecule, arising from the rotation around the second carbon-carbon bond (Newman projection must focus on one bond only). 2nd C has H, H and CH_3 as substituents, while 3rd C has H, H and CH_2CH_3 group as substituents. Convince yourself:

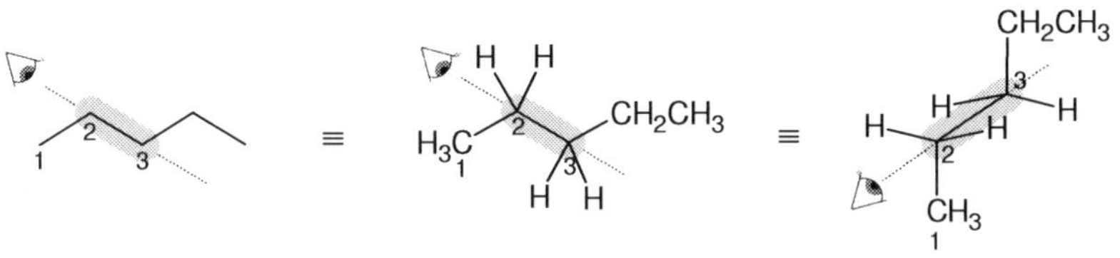

Fig. 11.7 A pentane molecule. The eye indicates how to look at it, in order to draw Newman projection along the C2–C3 bond.

Let us start with a Newman projection of staggered conformation, because firstly, it is easy to draw; secondly, it must be one of the special conformations.

Now use your imagination and rotate the back substituents in search of any new conformation, which is special in stability terms. I mean, any new conformation, which is characterized by the fact, that its stability is a *local minimum or a local maximum.* There are as much as four such conformations, when we consider rotation around the C2–C3 bond in pentane:

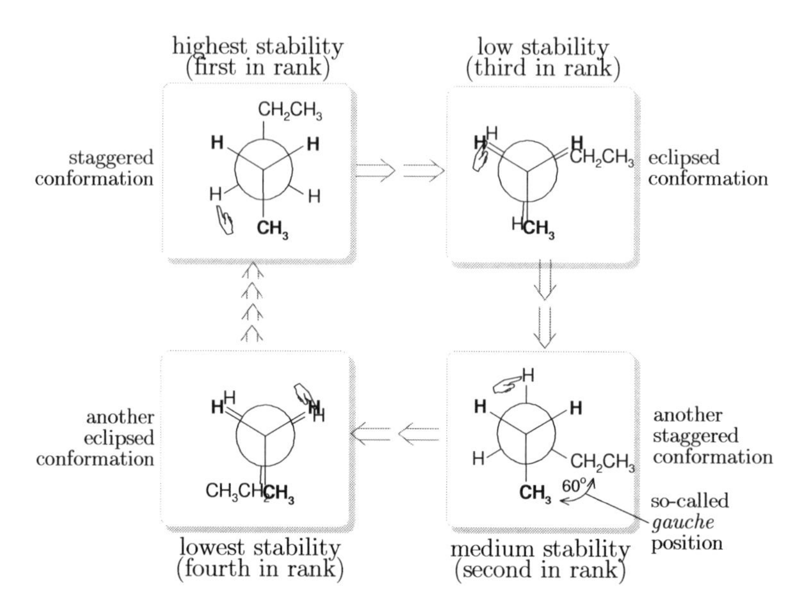

Fig. 11.8 Four special conformations of pentane, when rotated along the C2–C3 bond.

The first staggered conformation has the highest stability (CH_3 and CH_2CH_3 groups in a so-called **anti position**), while second staggered conformation is second in rank, because CH_3 and CH_2CH_3 groups are closer, and start to repel each other slightly (they are in a so-called **gauche position**; gauche position is whenever two substituents are separated by a 60° angle).

Eclipsed conformations have lower stabilities. Last Newman projection is an extreme case – eclipsed conformation with two large substituents just behind each other. Repulsion is the highest, so the stability is the lowest. Note that when you rotate back groups further from that point, you should not expect anything new. You are just regenerating conformations already met:

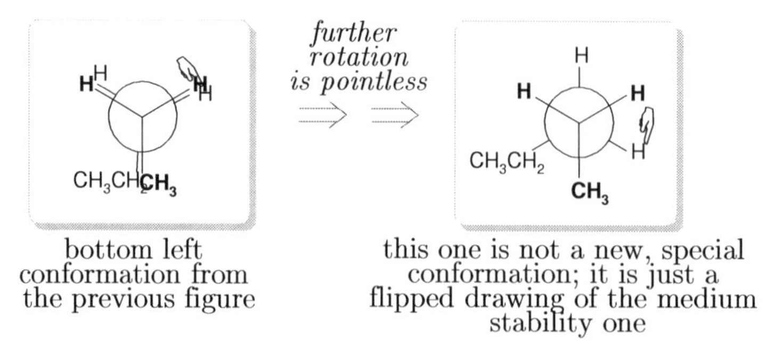

Fig. 11.9

Problems

Draw Newman projections of special conformations of the following molecules. Describe them as staggered/eclipsed, and find the most and the least stable ones:

11.1 rotation around C-C bond

11.2 rotation around C1-C2 bond

11.3 rotation around C-C bond

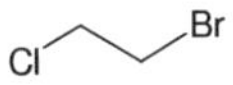

11.4 rotation around C3-C4 bond

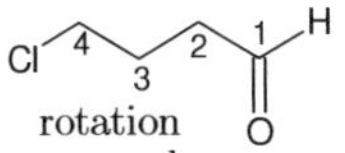

11.5 rotation around C-C bond

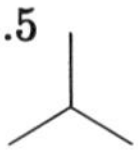

11.6 rotation around C2-C3 bond

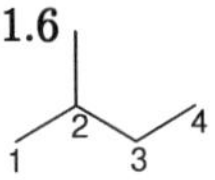

Consider pairs of Newman projections below and determine whether they are A, B, C, D, or E (only one of the answers is correct).

A: two identical conformations of one molecule, or
B: two different conformations of one molecule, or
C: two different, though isomeric molecules, or
D: two different molecules, which are not isomers of each other, or
E: two different Newman projections of one molecule.

11.7 and

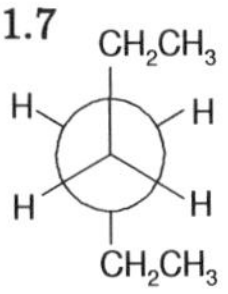

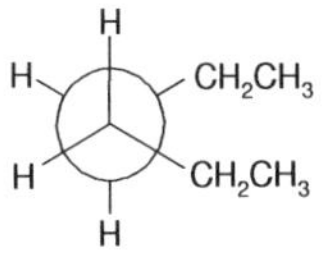

11.8 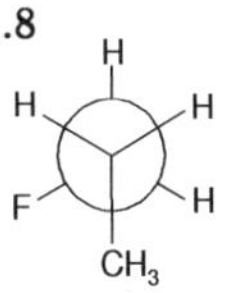 and

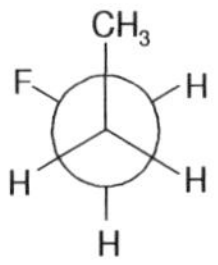

11.9 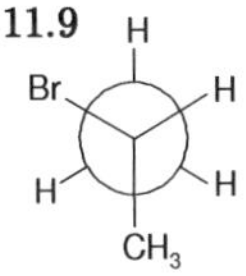 and

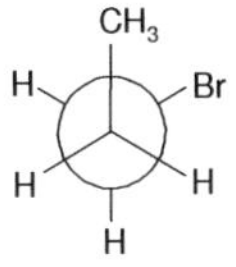

11.10 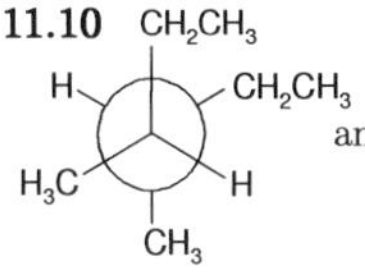 and

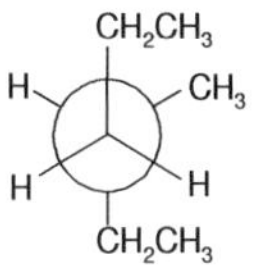

11.11 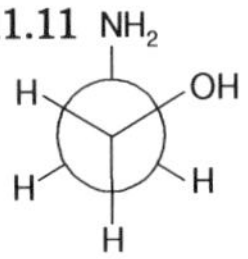 and

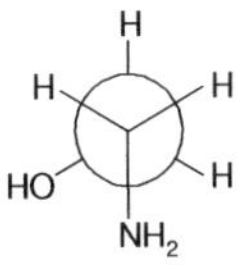

11.12 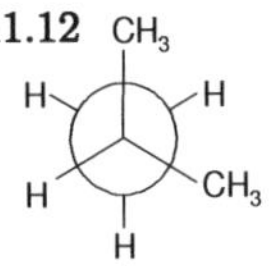 and

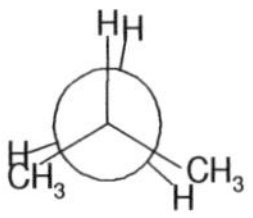

11.13 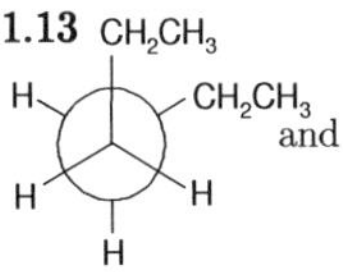 and

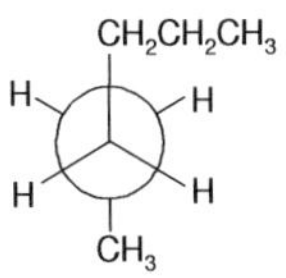

11.14 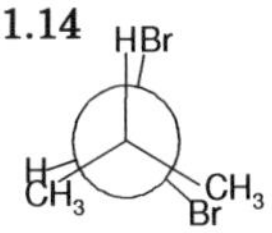 and

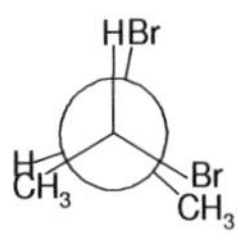

Change of conformation *vs.* bending bonds

Recall Fig. 4.6, where I drew 2-methylhexane few times, with chain bended different ways. Now, when you know that groups rotate around single bonds, it is clear why we can bend zigzags.

Simply, "straight zigzag" is not a frozen shape, rather one of many possible conformations. Below, I have drawn 2-methylhexane to show you how 3D rotation translates into 2D structure drawing:

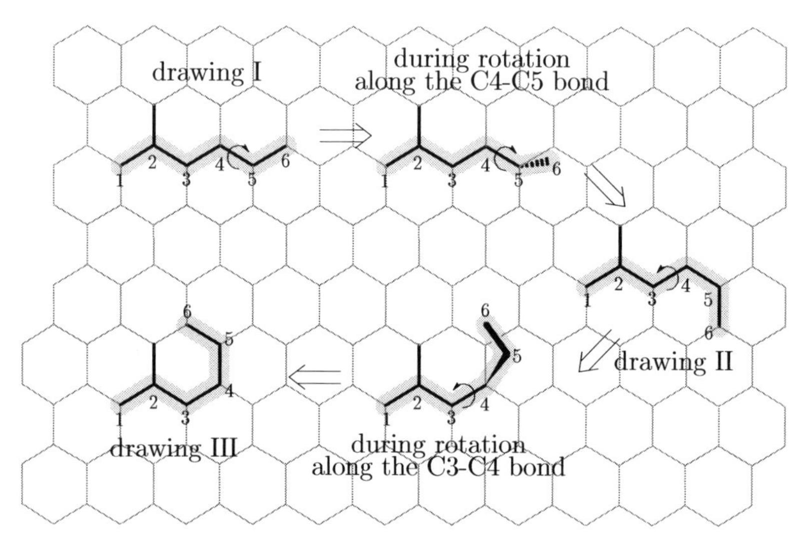

— stick depicts bond lying *on* the surface of paper
┅ dashes depict bond *under* the paper
◀ wedge depicts bond *above* the paper surface

Fig. 11.10

At the beginning, grab the 4th C atom with your left hand, and the 5th C atom with your right hand. Then, turn your right-hand fingers clockwise. C5–C6 bond goes *below the sheet of paper*, what we always illustrate on 2D drawings with a **dashed bond**. After 180°, the C5–C6 bond returns to the surface in a new place (drawing II), still however, fitting with the honeycomb.

Now, grab the 3rd and 4th bond. Rotate it. The entire CH_2CH_3 group goes *above the paper*, which we illustrate with a **wedge bond**, and returns to the surface in a new place after a 180° rotation.

We can refer to wedge and dash bonds as **stereobonds**, and we use them in stereochemistry, whenever there is a need to show schematically some 3D relations on simple, 2D drawings.

12 Conformations of rings and the ring strain

Chains are dynamic. Free rotation around single bonds causes constant changes of their overall shape. But what about rings?

Rings, just like chains, are not made of stone, and they are dynamic too. However, this subject requires separate treatment. In this chapter, I will introduce you briefly to conformations characteristic of small rings: three, four and five-membered. The next chapter will focus on conformations of cyclohexane.

In small rings bonds lie closer to each other than in chains, which gives rise to the so-called ring strain

Repulsion between bonding pairs sprouting from atoms, manifests itself in the fact that distances between bonds are maximized. This is a fundamental phenomenon, and atoms joined into a ring are no exception.

However, sp^3 atoms in molecules like cyclopropane, cyclobutane etc. do not adopt perfectly tetrahedral geometry. This is because in small cycles bonds are closer to each other, than in chains or larger rings (by design). Recall the drawing of a cyclopropane molecule – it is explicit that three internal angles (60°) are much smaller, than angles between C–C bonds in zigzags (~105°).

We cannot move away C–C bonding pairs in a cyclopropane, unless we break the cycle. The consequence is that there must be significant repulsion clenched inside. We refer to this inescapable repulsion in small rings as **ring strain**. Since a cyclopropane is the smallest, it experiences the highest ring strain, and is the least stable cycloalkane.

In cyclobutane angles between C–C bonds are larger. Therefore, the ring strain is smaller. In cyclopentane things get even more relaxed, while in cyclohexane there is no ring strain at all, because angles between C–C bonds are just like in chains. They are in favorable, perfectly tetrahedral arrangement.

Cycles larger than the cyclopropane, escape planarity to relieve the ring strain a bit

On professional 2D drawings, we draw cycles as flat, regular figures. However, in 3-dimensional reality, a planar ring is just one of the possible conformations. And more importantly, it is highly unstable.

The only cycle, which remains planar all the time, is three-membered cyclopropane. There is no other way to organize 3 atoms in space, you cannot rotate bonds or twist the molecule. In this regard, the cyclopropane is a unique molecule. It does not exhibit any conformational freedom.

However, in general, cycles adopt nonplanar conformations. It places their bonds a bit further away, so that the unfavorable repulsion is a bit relieved. Take a look:

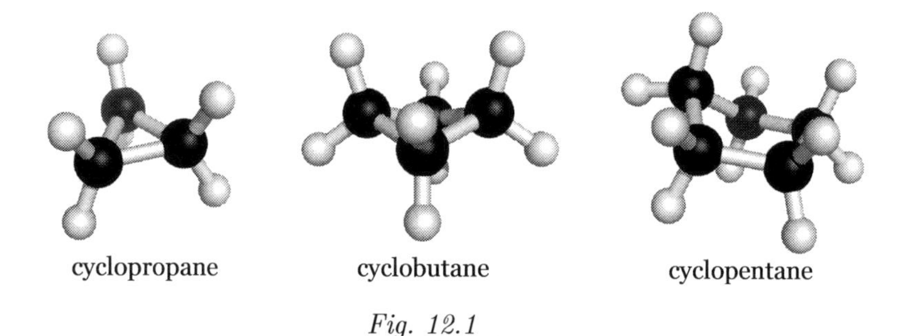

Fig. 12.1

Cyclobutane spends most of the time in such bended conformation, which some people call "a butterfly." You may think it is too poetic or exaggerated, but resemblance is not in the shape, rather in dynamic behavior. Cyclobutane exhibits conformational changes over time, by constantly interconverting. It flips side-to-side, like butterfly flapping its wings. We refer to such *conformational dynamism of rings* as **ring flipping**.

Planar conformation of cyclopentane is also highly unfavorable, and the molecule escapes it by pushing one of its Cs out of the plane. We call this favorite conformation an "envelope." The ring flipping in cyclopentane is a process, in which protruding C atom goes back to the mean plane, meanwhile one of neighboring CH_2 groups goes up.

The next member of the family is cyclohexane, but this guy has the entire chapter to himself.

13 Chair conformation of the cyclohexane ring

Cyclohexane is considered the most notable cycloalkane, because it is the most abundant non-aromatic ring in Nature. And there is a reason for that – cyclohexane has the least ring strain, and therefore is the most stable.

Of course, just like in case of cyclopropane and cyclobutane, a cyclohexane ring does not prefer to be in planar conformation. It should be no surprise to you if you remember a 3D picture of a menthol molecule from the first chapter (Fig. 1.1). Go back and watch it carefully.

Chair conformation of cyclohexane

The most favorable conformation of the cyclohexane ring, in all molecules containing it, looks like a deckchair and we call it **chair conformation** (this is exactly what computer software proposed while rendering the 3D model of menthol structure). Look at the chair:

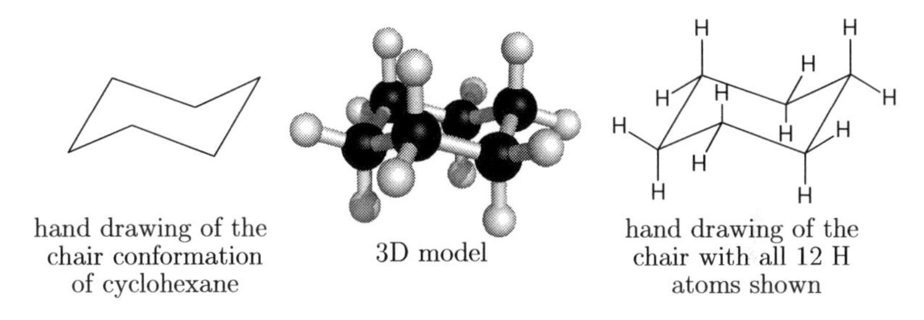

Fig. 13.1 Chair conformation of cyclohexane.

Some students hate it because its 3D structure is not so easy to comprehend, and in consequence the chair is hard to draw. However, I think we can handle such maladies in few steps. Step one:

Within the chair, all C atoms are identical and equivalent

Do not get deluded with how the chair looks – all six C atoms are identical, and they have identical surroundings in space. I know that at the first glance, it looks different, but that is only illusion, resulting from the fact that most brains, including mine, perceive the chair only in one way.

This false interpretation suggests that there are two Cs at two "tops" of a chair, and four Cs at its "sides." That is simply not true. Consider the following picture to see why:

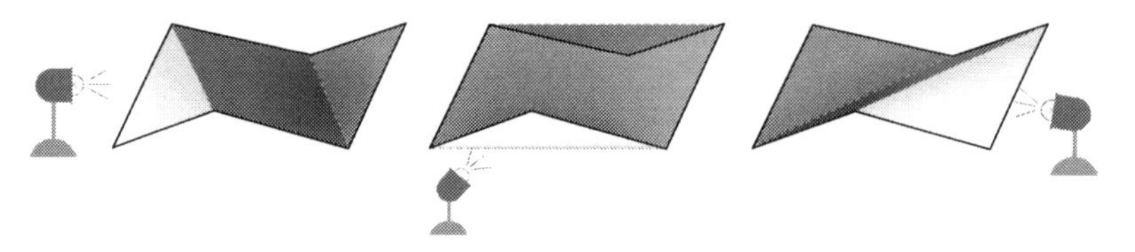

Fig. 13.2 Three equivalent ways to look at the chair.

I drew a chair, and copied it twice with ctrl+c. Follow black lines, which represent C–C bonds, to convince yourself that all three chairs are the same.

I have put three lamps to show, that you can illuminate the same chair from different directions, and your perception changes mysteriously. As you see, we can interpret every C as "top C," and in the same time, as "side C." Therefore, seeing tops and sides in the chair, is only an illusion resulting from the way your brain sees it. In fact, *all Cs are identical/equivalent.*

Although Cs are identical - substituents are not

We refer to positions and chemical bonds, which point vertically above and below the cyclohexane ring, as **axial**, while six bonds pointing diagonally around the cycle are **equatorial**. These two positions vary, because their spatial surroundings are different:

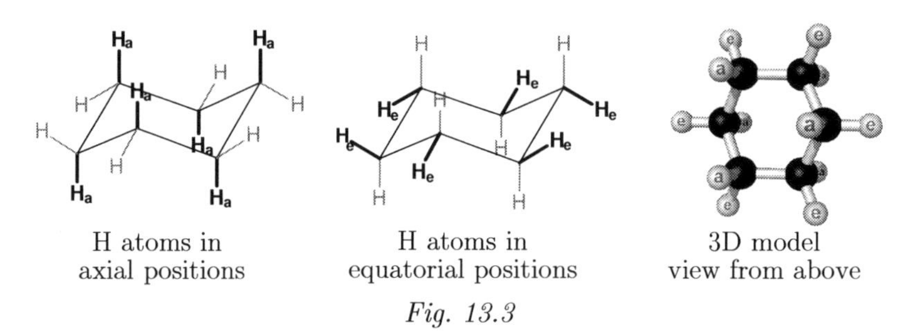

H atoms in axial positions

H atoms in equatorial positions

3D model view from above

Fig. 13.3

When atoms like H or any substituents are in axial positions, they are much closer to each other, than substituents in equatorial positions. It becomes apparent, when you realize that axial bonds are arranged in relation to each other, just like bonds in eclipsed conformations of chains. The most profound consequence is that *axial bonds and substituents experience higher repulsion.* On the other hand, equatorial bonds arrange like in staggered conformations of chains, and therefore, *equatorial bonds and substituents experience lower repulsion.*

Drawing chair conformation

Stereochemistry requires some talent. We must interpret flat pictures as if they were three-dimensional; handle such objects in imagination, and then draw them back on 2D paper. Some talented folks can draw the chair with twelve Hs without the slightest hesitation. But for many others it is an impressive achievement.

Here is an easy 6-step recipe, which I recommend to anyone who has trouble. Please, grab a pencil and practice on paper:

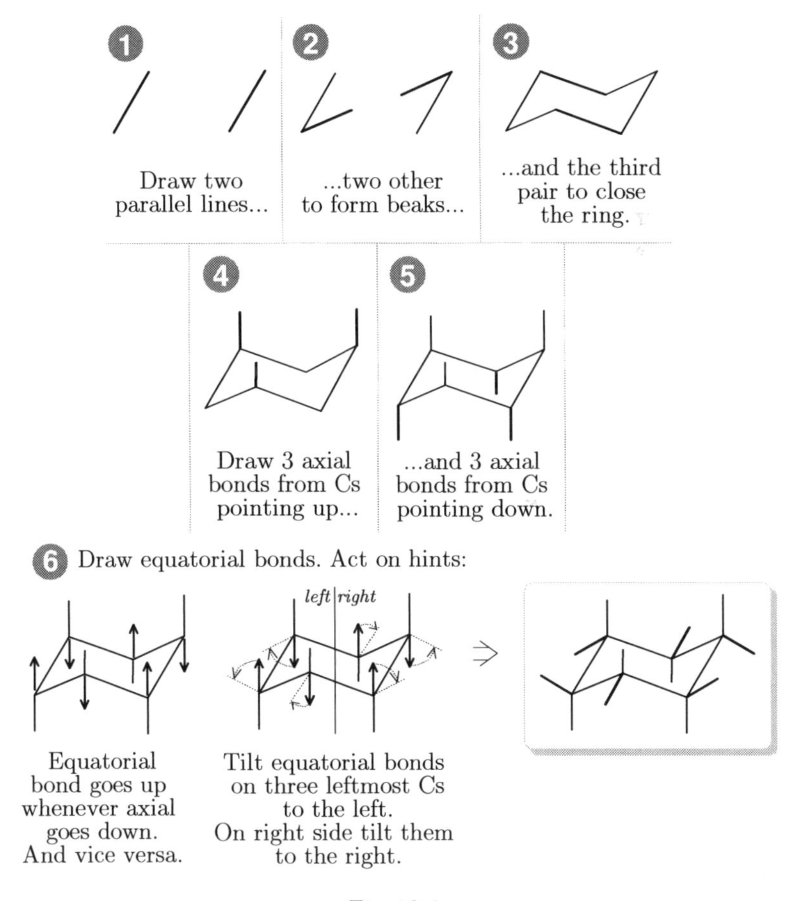

Fig 13.4

Ring flipping of the chair

Recall that cyclobutane flips side-to-side, and cyclopentane's "envelope" is dynamic too. The chair conformation of cyclohexane is not an exception, in this regard. It also exhibits **ring flipping**.

Consider the picture below to understand the process. Although all Cs are identical, I have marked them with numbers. It makes it easier to follow the changes:

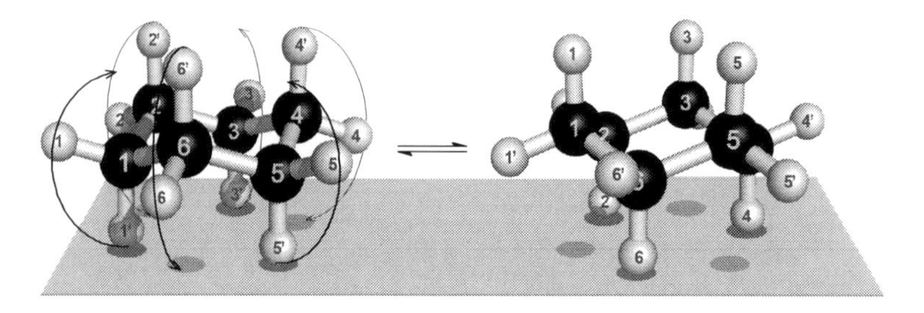

Fig. 13.5 Ring flipping in cyclohexane.

During the flip, upper CH_2 groups go down, while lower ones go up. Meanwhile, *every equatorial position changes into axial*, while *every axial becomes equatorial.*

Since the right chair in the above picture is a bit illegible, we had better redraw it after a slight clockwise rotation. Consider the flipping one more time:

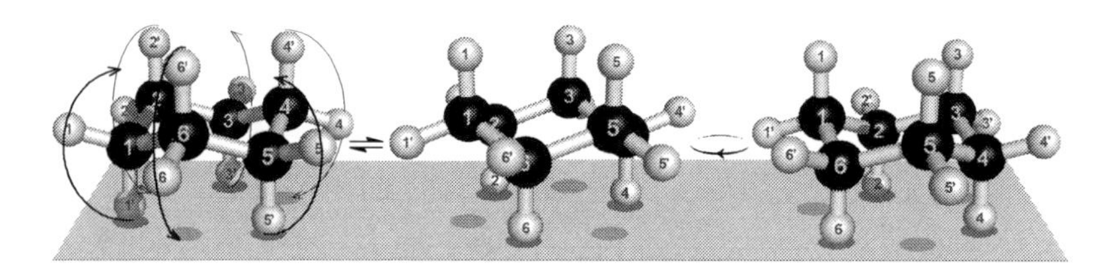

Fig. 13.6 Ring flipping in cyclohexane.

We always do hand drawings of the ring flipping with a slightly rotated flipped chair. Carefully compare the above picture with the one below:

Fig. 13.7 Hand drawn ring flipping.

There should be no problems with drawing the flipped chair. Simply, start with a pair of two differently slanted lines, and then follow the above-mentioned recipe. Moreover, remember that we always put a **double arrow** between both structures. It means "these two structures interconvert with each other; it works both ways with ease." In chemistry, we call such processes **reversible**.

An important thing to notice is that in case of cyclohexane, both *structures are identical; therefore, they are not two different chair conformations* of the cycloalkane. However, in molecules containing substituents attached to the ring, we will treat two chair conformations as different. They will differ in stability, just like various conformations of chains do.

When two chairs are different

Cyclohexane is just one of the millions of organic molecules. We have a better chance to spot this structural motif, as a part of the backbone of some larger molecules.

Methylcyclohexane is a relatively simple example. I drew it in form of a chair and then flipped it. Closer inspection reveals that both chairs are not identical. They are two different chair conformations of the molecule:

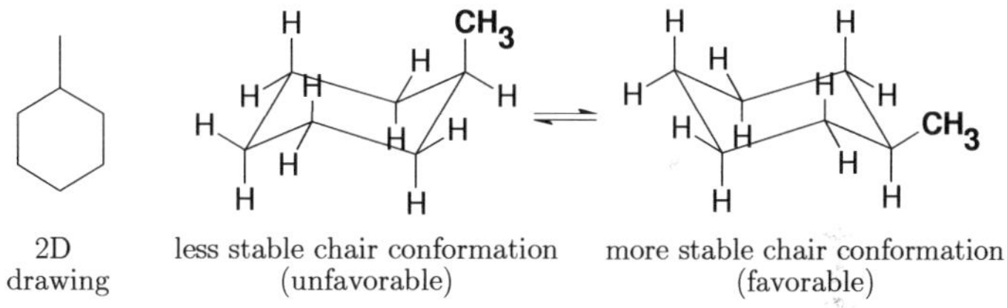

2D drawing | less stable chair conformation (unfavorable) | more stable chair conformation (favorable)

Fig. 13.8 Ring flipping in methylcyclohexane

Why is it so? In left conformation, a CH_3 group is in axial position, but after ring flipping it becomes equatorial. These are two different interconverting overall shapes of the molecule. Therefore, we must treat them as two different conformations.

And indeed, as you may expect, *they differ in stability* too. We know that substituents in axial position experience higher repulsion, because distances between them are smaller. Therefore, the left chair conformation is less stable, while the right, with a methyl group in equatorial position, is favorable. In consequence, a methylcylohexane molecule spends more time as "the right chair".

The larger the substituent the larger repulsive effect it will experience, when oriented axially. Therefore, we can expect stronger preference toward the chair with a large substituent pushed toward equatorial position. Indeed, in many molecules, the process of ring flipping is hindered, or even totally turned off.

Part IV STEREOISOMERS

14 *E*/*Z* stereoisomers and rules of priority

Free rotations around single bonds are the source of conformational changes in molecules. And what about multiple bonds? Imagine that you hold in hands a clay-matchstick model of ethene $CH_2{=}CH_2$. When you apply a force to rotate CH_2 groups, clay balls joined with two matchsticks would rather smash than rotate. *In molecules, there is no rotation around double or triple bonds.* In case of C≡C, it does not mean anything special, but prohibited rotation around C=C draws a new story to light.

Take a look at two pictures of but-2-ene below. Certainly, there are plenty of different conformations possible to draw, because of rotation around C–C single bonds. But are these two structures two conformations of one molecule? No, simply because there is no way to rotate groups around a double bond. You cannot rotate half of a model to obtain the second. And molecules also do not experience such a process:

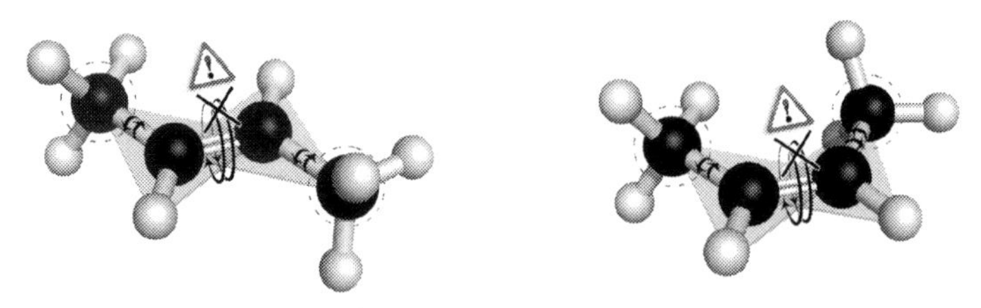

Fig. 14.1 Two distinctive, inconvertible but-2-enes.

Therefore, these two structures are not interchangeable, transient shapes. Instead, both models show *two different molecules.* You can even buy one or the other as separate, pure substances in a chemical store, as long as the price about $ 2300 per pound does not put you off.

But how can two distinctive molecules have the same IUPAC name, while names are supposed to be unique IDs of molecules? Well, they do not have the same name, actually. Recall that in Chapter 9, where I listed steps we go through to name a molecule, I wrote: "Whenever applicable, add stereodescriptors at the very beginning of IUPAC name." And the time has come.

Stereodescriptors are additional chunks of IUPAC name designed to differentiate them in special occasions, just like the one with but-2-enes. As you will soon see, we will use *E* stereodescriptor for the left molecule, while the right alkene will be dubbed into *Z*. However, before we get to details, let us familiarize with a new term, which is the leitmotif of this part of the book:

Stereoisomerism

You are ready to learn that there are two types of isomerism in organic chemistry: **structural isomerism** and **stereoisomerism**.

Menthol and citronellol were examples of structural isomers. Structural isomers vary in connectivity of atoms, and therefore their structures can be so strikingly dissimilar, that isomeric relation becomes hard to notice. We have to count atoms meticulously to check the atomic composition, which may be the only thing in common.

On the other hand, we have pairs of molecules like two but-2-enes. They are stereoisomers. A characteristic feature of stereoisomers is that they *do not* differ in connectivity of atoms. Ironically, their structures are so strikingly similar, that isomeric relation might be hard to notice. We have to make an effort to spot the difference.

All isomers are molecules made of the same set of atoms. General definition of isomerism says so. In case of two but-2-enes it is C_4H_8, there is no doubt. Nevertheless, quite contrary to structurally isomeric molecules, molecules that are stereoisomers of each other, share much more than just atomic composition.

In fact, they are made of the same set of groups – not only the same set of atoms. *The only difference between stereoisomers is in the fixed locations of these groups in space.*

Look, both *E* and *Z* but-2-ene are made of the same CH=CH fragment, and two CH_3 groups. Nevertheless, they differ because there are two fixed, unchangeable ways to locate these groups in space. We refer to such possible arrangements as different...

Configurations of C=C bonds

Fixed configurations result from prohibited rotation around double bond. For that reason, we have many similar alkenes, which are in fact pairs of different stereoisomeric molecules: *E* and *Z*.

You must be careful now, when you draw structures with double carbon-carbon bonds, because they clearly indicate which configuration – *E* or *Z* – you have in mind:

Fig. 14.2

Two stereoisomeric but-2-enes differing in the configuration of the double bond.

In addition, prohibited rotation around double bond means, that although you can feel free to bend single bonds in chains, it is forbidden when the bond is attached to C=C.

How to assign *E* or *Z* stereodescriptors?

1 Look at the first carbon atom in the C=C bond. It has two single bonds to two different atoms. Compare their weights. The heavier one "wins."

2 Look at the second C in C=C bond. It also has two single bonds to two different atoms. The heavier one "wins."

3 Write one of two stereodescriptors:
E – when winners are on opposite sides of C=C, or
Z – when winners are on the same side of C=C.

In case you do not know: weights of atoms grow in the periodic table from left to right, and from top to bottom, so the sequence is as follows (H loses all, iodine always wins):

lightest **H < C < N < O < F < Cl < Br < I** heaviest

Officially, this ranking represents the **priority of substituents** (we will delve into details shortly).

Let us take a look how it works for the left but-2-ene stereoisomer. The first C in C=C has a single bond to **C** (in CH_3 group) and H. **C** is heavier, so it wins. The winning substituent – CH_3 group – points *upwards*.

The second C in C=C has single bonds to H and **C** (in CH_3). Once again **C** wins, but the winning substituent points *downward*. Since both winners are *on the opposite sides* of C=C we assign *E* stereodescriptor to this particular configuration.

The right but-2-ene will be a *Z* stereoisomer, because winning substituents are on the same side of C=C (by the way, this might be some help for you with remembering the meaning of *E* and *Z*: namely, *E* comes from German *entgegen* meaning "against," while *Z* from *zusammen* meaning "together").

The configuration must be implanted in IUPAC name

The name of a molecule must be its unique ID. It is not the case, when we look at the word "but-2-ene." It is ambiguous, because it describes two different stereoisomeric molecules.

Therefore, we should implant stereodescriptors into IUPAC name. Rules of nomenclature say that we add them on the very beginning. It goes with the locant of a double bond, but without a hyphen: e.g., 2*E*, 2*Z*. Moreover, we enclose such 2*E* or 2*Z* in parentheses:

(2*E*)-but-2-ene (2*Z*)-but-2-ene

Fig. 14.3

However, hiding the locant is very popular. It does not introduce any ambiguity: there is only one (*E*)-but-2-ene and only one (*Z*)-but-2-ene to draw. Omitting the locant of *E* or *Z* stereodescriptor is against IUPAC recommendation ("you can hide the locant only when it is omitted in the very name too; otherwise, repeat it"), but the truth is that no one cares.

Assigning *E*/*Z* in harder cases

In but-2-enes assignment of stereodescriptors is a trifle, because we have to compare H and C (in CH_3), so you recognize the winner at the first glance.

But how can we concede victory when the two competing substituents start with the same atom? In such situations, the fight will not be decided in the first round. We go deeper into the structure of both substituents, drawing them as "trees."

Let us ponder on the fight, which takes place on the left C atom of the C=C bond. At first, we have **C** (in **CH_3**) *vs.* **C** (in **CH_2OH**). Therefore, we cannot declare the winner.

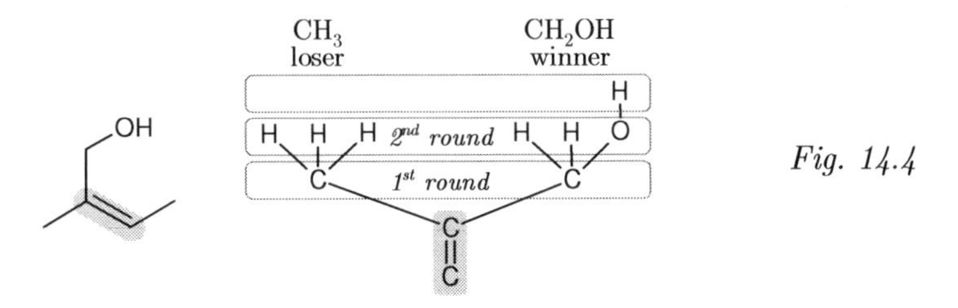

Fig. 14.4

The show must go on. In the second round, we compare atoms deeper in the structure of both substituents. As the "tree" shows, CH_3 puts three H atoms to the test, while CH_2OH has H, H, and O. The case is clear: CH_2OH group beats CH_3, because O is heavier than any of 2nd layer atoms in the CH_3 group. The winner, CH_2OH group, is on the same side as the winner of the second fight (passed over), so the configuration is determined as *Z*:

winner OH

Z

loser winner

loser

Fig. 14.5

There is a special rule of the game, which says that *one heavier atom is enough to beat endless number of others.* For example, a CH_2Br group wins CCl_3, although the latter weights more as a whole. We do not care in wholes. Br wins with an infinite number of Cls, because fights are one by one, and one on one:

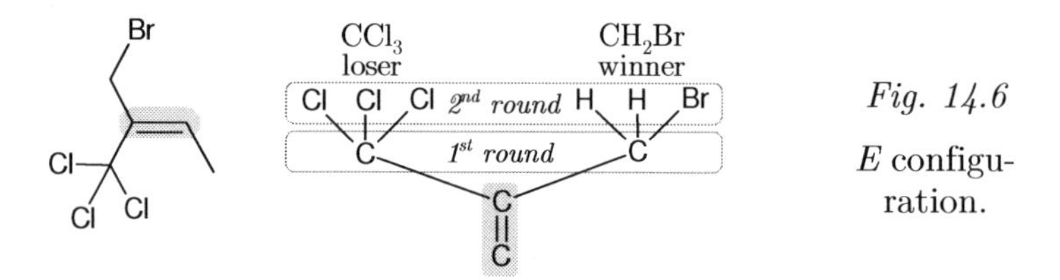

Fig. 14.6

E configuration.

Second special rule of the game says that *double/triple bond to some atom Y counts as two/three single bonds to two/three Ys.* Therefore, "tree" representation of a C≡N group is a C atom with three single bonds to three Ns. This rule makes a nitrile group the winner in the fight with CH_2NH_2:

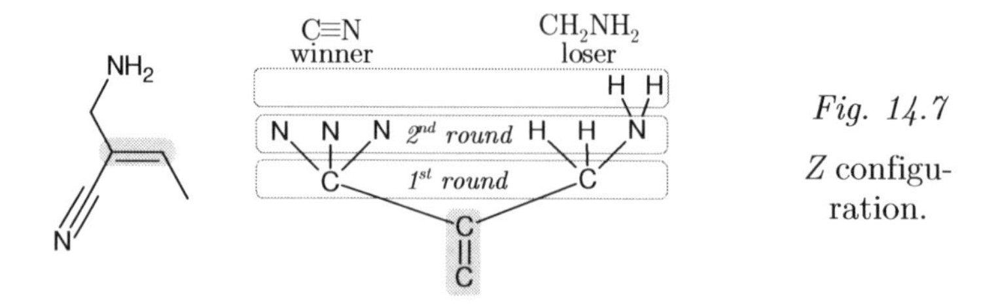

Fig. 14.7

Z configuration.

Officially speaking, the game determines **priority of substituents**, and we refer to the winning one as *the highest priority substituent.* Students confuse "*this* priority" with a priority of functional groups in general IUPAC nomenclature (recall tables in Chapters 6 and 7).

Remember that we use priority rules described here only when we assign stereodescriptors (E and Z; in the future, we will also use it while considering R and S stereodescriptors).

Problems

For each of the following pairs decide which of the substituents wins the fight (which of them has higher priority):

14.1 *vs.*

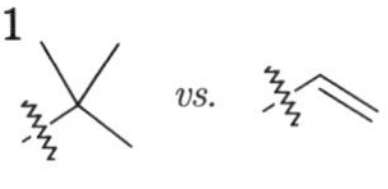

14.2 *vs.*

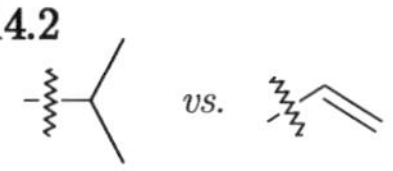

14.3 Ph, N, Ph *vs.* OH

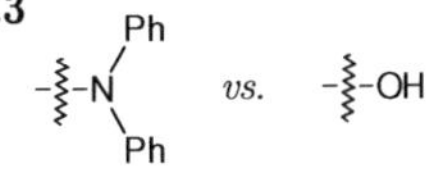

14.4 *vs.*

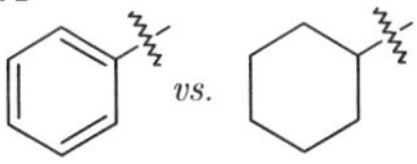

14.5 *vs.* Br

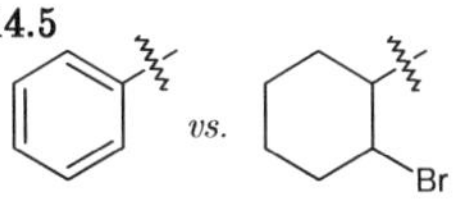

14.6 *vs.*

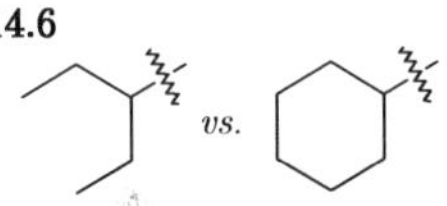

14.7 *vs.*

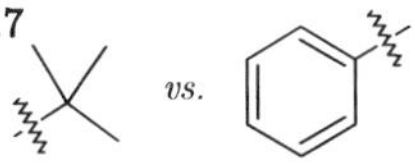

14.8 Cl *vs.* Cl

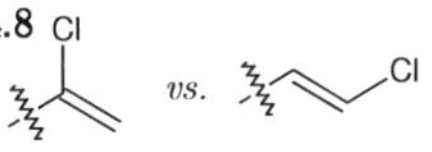

14.9 *vs.* F

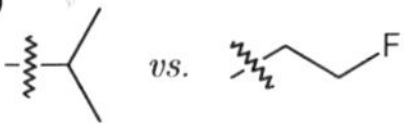

14.10 N *vs.* OH

14.11 OH *vs.* OH, O

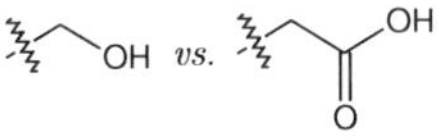

14.12 OH, O *vs.* OH, O

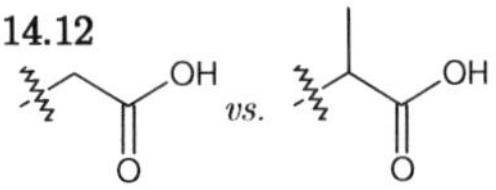

In the groups below order substituents according to decreasing priority (from the best of the best, to the worst looser):

14.13 $-CH_2Cl$, $-Br$, $-NCH_3$, $-CH_2NH_2$, $-CH_2Ph$, $-CH_2F$

14.14 $-CH_2Cl$, $-CHFCH_3$, $-CH_2CH_2F$, $-NHCH_3$, $-CH_2OH$, $-CH_2Ph$, $-CH_2F$

14.15 $-COCH_3$, $-OCH_3$, $-CONH_2$, $-NH_2$, $-CH_3$, $-CH(CH_3)_2$

14.16 $-COOH$, $-C{\equiv}CH$, $-CH{=}CH_2$, $-COCH_3$, $-CH(CH_3)_2$, $-C(CH_3)_3$

14.17 $-C{\equiv}N$, $-CH_2OH$, $-CH_2CH_2OH$, $-OCH_3$, $-OC(CH_3)_3$, $-CH_2C(CH_3)_3$

Use your experience in judging fights and determine E/Z configurations of double bonds in stereoisomeric molecules below ('Ph' = monosubstituted benzene ring):

14.18 Ph

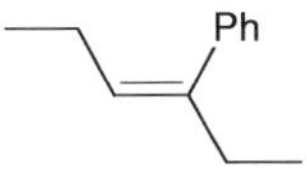

14.19 Ph

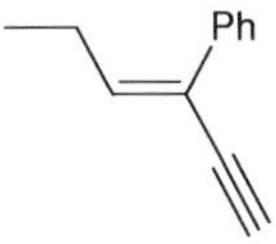

14.20 Ph

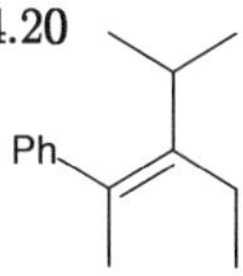

14.21 14.22 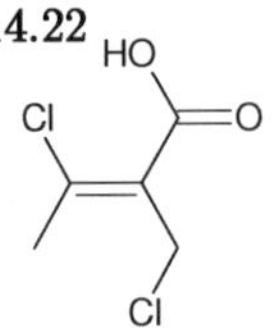 14.23

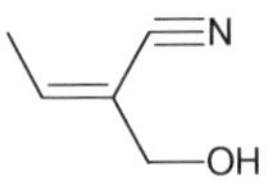

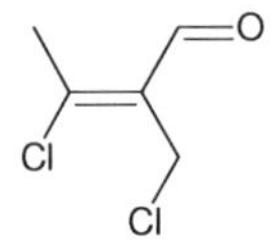

What if a molecule has more than one C=C bond?

E and *Z* descriptors are determined for each of C=C bonds, obviously. What is more important is that in such situations we have more than two stereoisomers. For example, with two C=C bonds in a molecule, there may be as many as four stereoisomeric molecules: *E*,*E*; *Z*,*Z*; *E*,*Z* and *Z*,*E*:

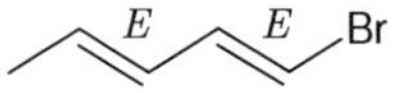

(1*E*,3*E*)-1-bromopenta-1,3-diene 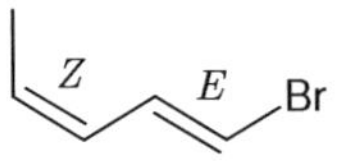 (1*E*,3*Z*)-1-bromopenta-1,3-diene

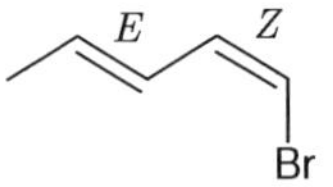

(1*Z*,3*E*)-1-bromopenta-1,3-diene 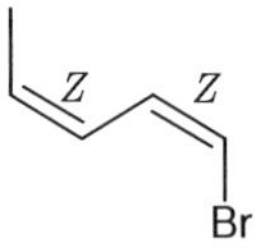 (1*Z*,3*Z*)-1-bromopenta-1,3-diene

Fig. 14.8

C=C bonds for which two configurations do not exist

There are molecules which do not exhibit *E*/*Z* stereoisomerism, although they have a C=C bond inside. It happens whenever there are *two identical substituents on at least one C atom of a double bond.* That is the case in a citronellol molecule, where we have two CH_3 groups at the end of a chain. Look at other examples too:

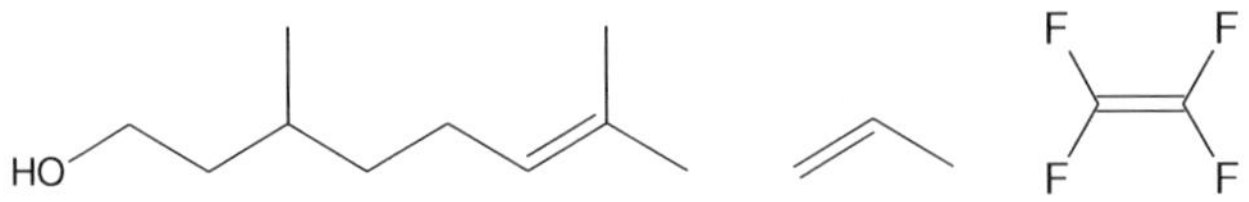

Fig. 14.9 Molecules which do not exhibit *E*/*Z* isomerism.

In propene (*not* prop-1-ene!), there are two H atoms at the end. The terminal C=CH_2 group never exhibits *E*/*Z* stereoisomerism. The situation is even more clear in tetrafluoroethene (*not* 1,1,2,2-tetrafluoroeth-1-ene!), which is a substance from which we make **Teflon** for skillets.

The reason for above molecules not to exist as pair of E and Z isomers is purely geometrical – new drawing with interchanged positions of substituents at any C atom of C=C bond, will still be an identical molecule. Try this on paper.

Problems

Assign E or Z, if applicable:

14.24

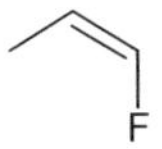

F

14.25

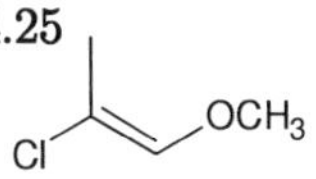

Cl OCH$_3$

14.26

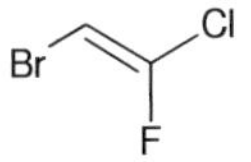

Br Cl F

14.27

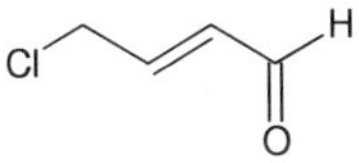

Cl H O

14.28

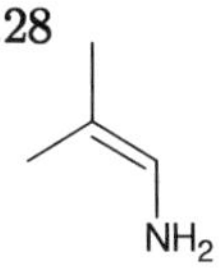

NH$_2$

14.29

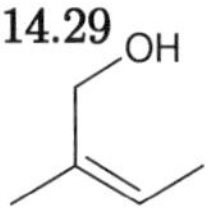

OH

14.30-14.35 There are 6 isomeric acyclic alkenes with a formula C_5H_{10}. Draw all of them, and assign E/Z stereodescriptors wherever applicable. Finish the task writing full IUPAC names of all these molecules.

Never confuse configurations with conformations

It is a pity that both terms sound so similar. Some students get lost in the maze of all these new words.

Two structures can vary in configurations of a C=C bond, which are fixed and inconvertible. When two drawings differ in configurations, they depict two different, but stereoisomeric molecules.

On the other hand, one molecule can have from 1 to endless number of conformations. They all are just transient shapes, which molecule changes freely through bond rotations and/or ring flipping. Different drawings of conformations still depict one molecule, which reshapes itself over time.

15 *Cis/trans* descriptors and their applications

As an alternative to *Z* and *E* stereodescriptors, we may sometimes use Latin words: *cis* and *trans*. Nevertheless, there is a snag.

While *E*/*Z* system is universal (applicable to all alkenes), the *cis*/*trans* system is not. This is because *E* and *Z* denote configurations of C=C, while *cis* and *trans* descriptors *tell the relation of two and only two substituents.*

Therefore, we can apply *cis* and *trans* descriptors only to alkenes with CH=CH double bond, and two substituents attached. *Cis* means they are "on the same side," while *trans* means "on the other side" (therefore, *cis* acts like *Z*, and *trans* is like *E*):

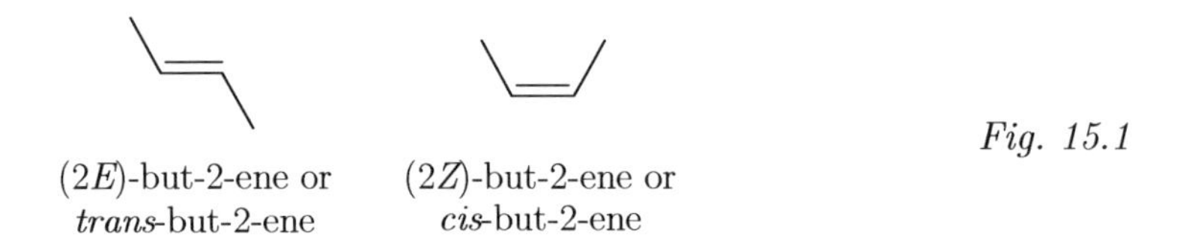

Fig. 15.1

Determining *cis* or *trans* relation in alkenes with more than two substituents is impossible. However, even in case of a disubstituted C=C bond, it is recommended nowadays to treat *E*/*Z* system as superior.

Problem

15.1 Comb through molecules **14.24-14.29**, and assign *cis*/*trans* descriptors wherever applicable.

Rotation of bonds in rings is prohibited

The last chapter was devoted to the fact that prohibited rotation around C=C bonds introduces stereoisomerism. There is *no way to rotate bonds in rings* either, so there is another type of stereoisomerism among cyclic molecules.

Take a look at any of models below. Can you rotate C–C single bonds just like we did for chains? No way. If needed, once again imagine you hold a model made of modeling clay and matchsticks, to convince yourself:

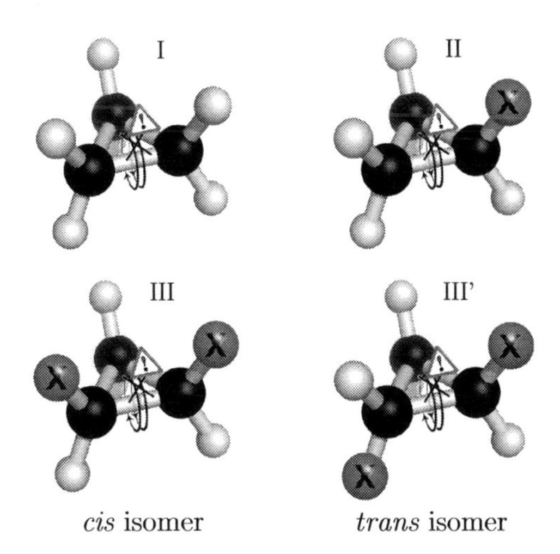

Fig. 15.2 I – cyclopropane, II – monosubstituted cyclopropane, III & III' – two disubstituted cyclopropanes: *cis* and *trans.*

Consider models **III** and **III'** – these are two distinctive disubstituted derivatives of cyclopropane. Prohibited rotation around bonds in rings generates the possibility to have *two different arrangements of substituents in relation to each other.* Obviously, these arrangements are fixed and not interchangeable. Therefore, model **III** and **III'** are different, though stereoisomeric molecules. We can describe them as *cis* isomer (substituents on the same side of the ring), and *trans* (opposite sides of the ring).

Whenever there is only one substituent, like in model **II**, there is no place for stereoisomerism. Test it yourself: put the X substituent in various places to see that each time you end up with another drawing of the same molecule. Stereoisomerism arises only *when the ring has two or more substituents.*

Cis and *trans* descriptors are applicable to disubstituted rings only

For now, you need to know that *cis* and *trans* descriptors cope well with substituted rings, but on one condition (the same as before): it must be the molecule with exactly two substituents attached to the ring.

Using it in relation to rings with more than two attachments is impossible. In such molecules, we use R/S system of stereodescriptors. It is universal just like E/Z system, but we leave it for later.

Cis and *trans* relations on the chair conformation

We know that cyclohexane spends most of its time in chair conformation. Since *cis* and *trans* descriptors relate to sides of the ring, we need to understand where these sides are, when we look at the chair representation. The point is that one side is "above," and the other is "below" the so-called mean plane of the ring. This mean plane would contain the entire ring, if it adopted totally planar conformation.

Some axial and some equatorial bonds go up, while others go down. The direction is clearly visible in case of axial bonds. Equatorial ones, however, require more caution:

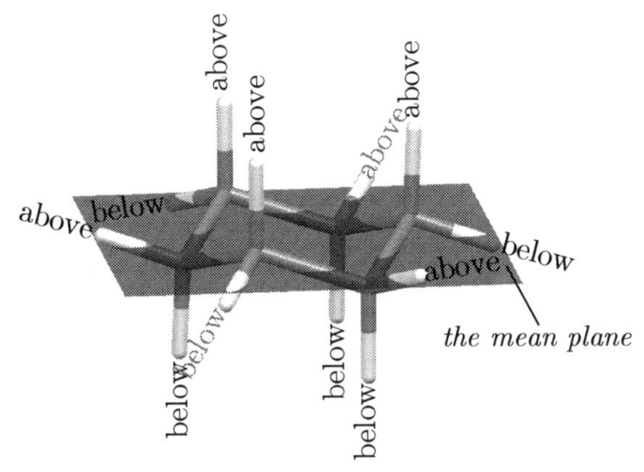

Fig. 15.3 Two sides of the chair: above and below the mean plane.

When one substituent is located above, while the second below the mean plane, they are on opposite sides of the ring, that is to say, in *trans* relation. When both substituents are either above or below, they are on the same side of the ring, and we deal with a *cis* isomer.

And how to show *cis* and *trans* relations on simple 2D structure drawings?

On 2D drawings, we can easily show stereoisomeric relations between substituents using **wedge and dash bonds**, which you have already seen in Chapter 11. All of the bonds in the ring lie on the surface of paper. Substituent going up is drawn with a wedge, while the one below the paper has a dashed bond. It really works, and you will see that it is easy to translate it into chairs.

Consider the example, 4-methylcyclohexan-1-ol, which in fact exists in two stereoisomeric versions. Here come drawings of *cis*-4-methylcylohexan-1-ol. You can draw them with two wedges or two dashed bonds. It does not matter; both are equivalent:

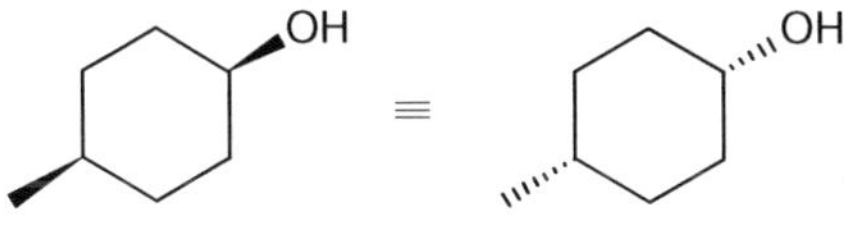

cis-4-methylcyclohexan-1-ol

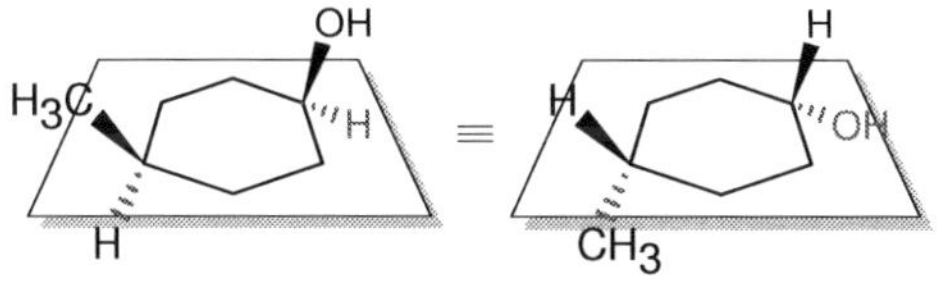

Fig. 15.4

Some people like to interpret drawings with wedge and dash bonds by tilting the paper, so I have drawn how it should look in your imagination. And this is *trans* isomer:

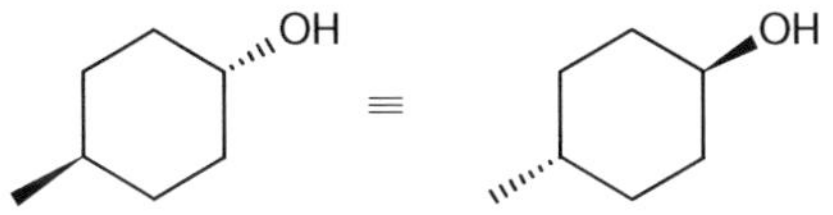

trans-4-methylcyclohexan-1-ol

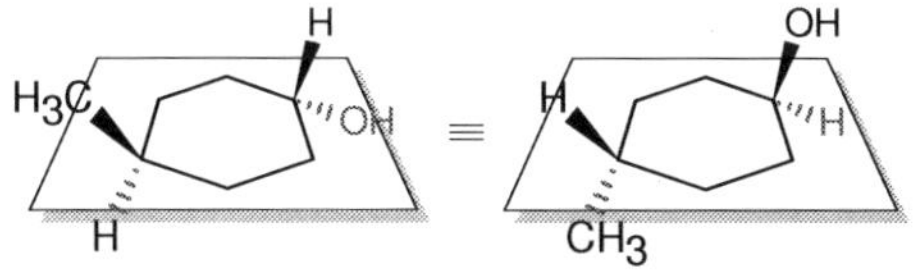

Fig. 15.5

Problems

15.2-15.8 Draw 2D structures of 7 isomeric dimethylcyclohexanes using wedge and dash bonds, wherever applicable. Write their IUPAC names.

Incorporating information about *cis/trans* relations between substituents in disubstituted rings is really easy, as you see. Let us now try to draw all these, using chairs.

Drawing chair conformations of *cis/trans* stereoisomers

1 Draw the chair.

2 Attach one substituent to whichever C of the ring – all of them are equivalent, after all. You are free to choose between axial and equatorial positions. We are going to flip the ring, so the substituent will turn into a second possible version, anyway.

3 Attach second substituent, but this time:

- localize correct C, in order that the distance between both substituents is correctly reproduced,

- decide whether it should be attached via axial or an equatorial bond, so that you correctly reproduce the spatial relation (*cis* or *trans*).

The picture below shows chair conformations of a last seen molecule: *trans*-4-methylcylohexan-1-ol. Having the chair drawn, the first thing to do is to put first substituent. I have picked a hydroxyl group, and put it on an axial bond of the "rightmost" C. This C got locant "1" by default. Afterwards, I localized C number "4" and attached CH_3 group via an axial bond, in order to place both substituents in correct, *trans* relation, and eventually, draw the flipped chair:

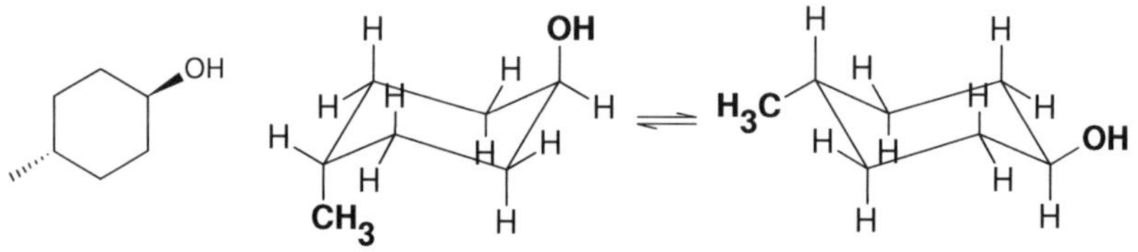

Fig. 15.6 Two different chair conformations of the *trans*-4-methylcyclohexan-1-ol molecule.

Interestingly, two chair conformations of our *trans* stereoisomer differ dramatically in geometric and stability terms. In left one, both substituents are in axial positions, which is, as you should remember unfavorable. Substituents in axial positions experience stronger repulsion from other axial "things," just like in eclipsed conformations of chains.

On the other hand, the right chair has both substituents in equatorial positions. That must be dramatically more stable, because distances between equatorial bonds are larger. The conclusion is clear: *trans*-4-methylcylohexan-1-ol stays most of the time in the form of the right chair, and the whole ring flipping is strongly hindered.

Now let us see what is going on in the *cis*-4-methylcyclohexan-1-ol molecule. This is how its conformational dynamism looks like:

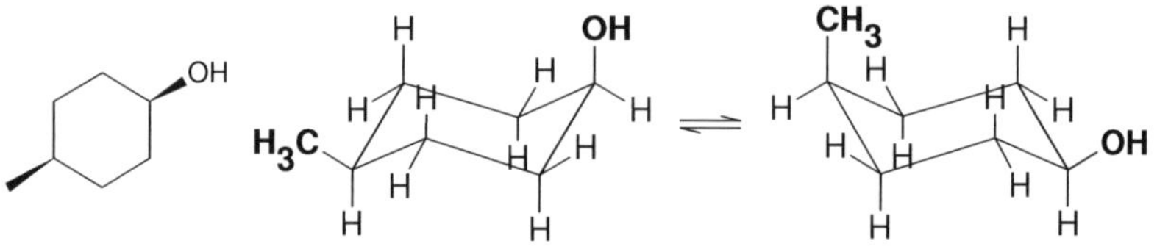

Fig. 15.7 Two different chair conformations of *cis*-4-methylcyclohexan-1-ol molecule.

Which conformation is more stable? The question is harder than before, because both chairs have one substituent in axial, and one in the equatorial position. The difference between chairs is less apparent. Ask yourself: is chair feeling more stable, when OH is axial, or when CH_3 is axial? That is the snag.

We need to realize, that CH_3 group is larger, and occupies more space. Therefore, it feels (and exerts!) more repulsive forces onto its surroundings, than smaller group would do. The answer slips out now.

If there is no other way, than to have a substituent in the axial position all the time, it is better to put the smaller one out there. The chair with an axial OH group is more stable, so the *cis*-4-methylcyclohexan-1-ol molecule spends a bit more time with a methyl group pushed to the equatorial position.

Problem

15.9 Can you predict which stereoisomeric molecule, as a whole, is more stable:

cis-4-methylcylohexan-1-ol or *trans*-4-methylcylohexan-1-ol?

16 Asymmetric carbon atoms

In this chapter, we are going to take a rest from stereoisomerism, and focus on the new concept, which is very important, yet easy-to-grasp. It is another feature of molecules, which we can notice only when looking at them in three dimensions.

Many organic molecules contain a so-called **asymmetric carbon atom** in their structures. The very idea is simple: *any C sp^3 atom, which sprouts its four single bonds to four structurally different substituents, is the said asymmetric carbon atom.* I devote the rest of this part of the book to asymmetric carbon atoms, and consequences of their existence.

Finding asymmetric carbon atoms on the structure drawings

From now, I will refer to such Cs as C*. This asterisk will also be used as a marker on structure drawings.

Convince yourself you got the idea. Finding C* in molecules is easy – we just look for any C atom with four dissimilar substituents. Many of them were already on the drawings in the book. These are other examples (note that undrawn H atoms count as one of four nonidentical substituents on C*):

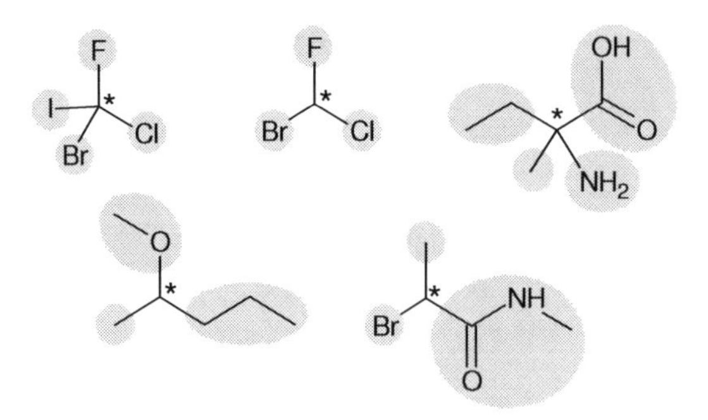

Fig. 16.1 All of these molecules have one C*.

In the following problems, I ask you to find C*s in various molecules. A useful hint for now and later, is that the condition of having *four* different substituents, automatically excludes C atoms in ubiquitous CH_2 and CH_3 groups from the crowd of suspects.

Furthermore, do not waste your time to evaluate C atoms with multiple bonds (C sp^2 or sp hybrids). By design, they bond less than four substituents.

Problems

In the following molecules, mark all asymmetric carbon atoms with an asterisk. Do not act routinely and take into account, that molecule may not have C*, while another may have more than one.

16.1 O Cl

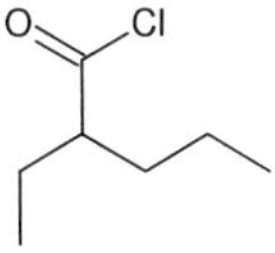

16.2 Cl OH

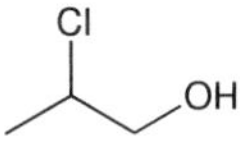

16.3 O

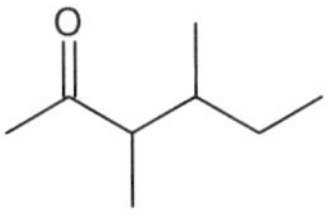

16.4

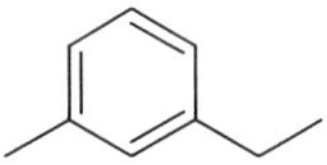

16.5 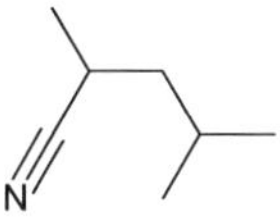 N

16.6 F Cl

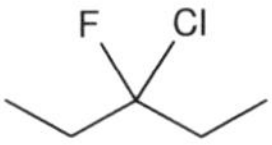

16.7 N

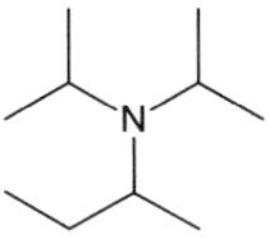

16.8 Br

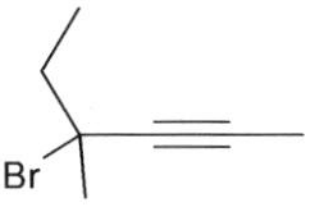

16.9 Br O N

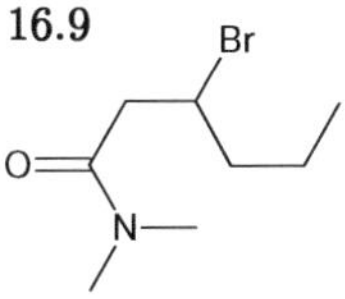

16.10 N

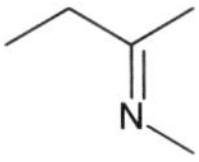

16.11 O O

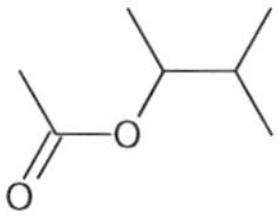

16.12 O

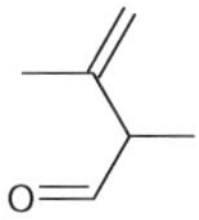

16.13 OH OH OH

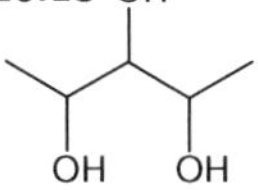

16.14

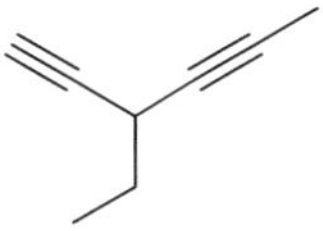

16.15 O O

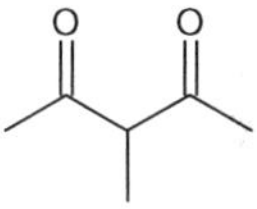

Finding C* among atoms in rings

When you evaluate a C sp^3 atom, which is a member of a ring, treat two halves of the ring – two sides with C sp^3 in question lying on the border – as two separate substituents. Molecules below are examples, where both halves are the same:

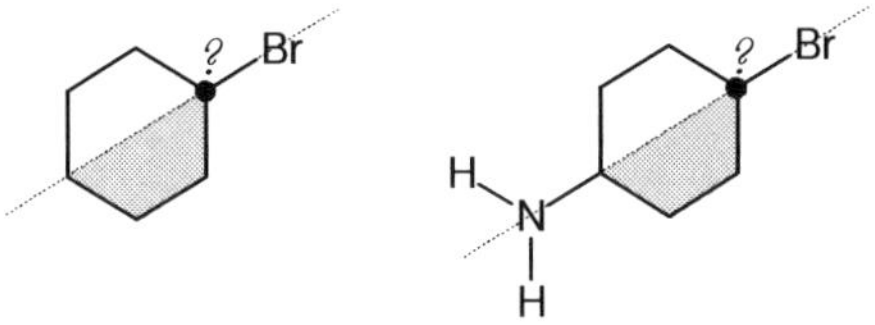

? Br

H N H ? Br

Fig. 16.2

Since they turn out to be identical, we cannot classify C sp^3 atom in question as asymmetric. For the same reason another suspect – the C atom bonded with NH_2 – is not a C*.

On the other hand, whenever there is a difference to spot, we treat both sides of the ring as two distinctive substituents. Take these two molecules as examples:

Br and: Br — two C* atoms in 2-bromocyclohexan-1-amine
NH_2 NH_2

Br and: Br — two C* atoms in 3-bromocyclohexan-1-amine
NH_2 NH_2

Fig. 16.3

Problems

In the following molecules mark all asymmetric carbon atoms:

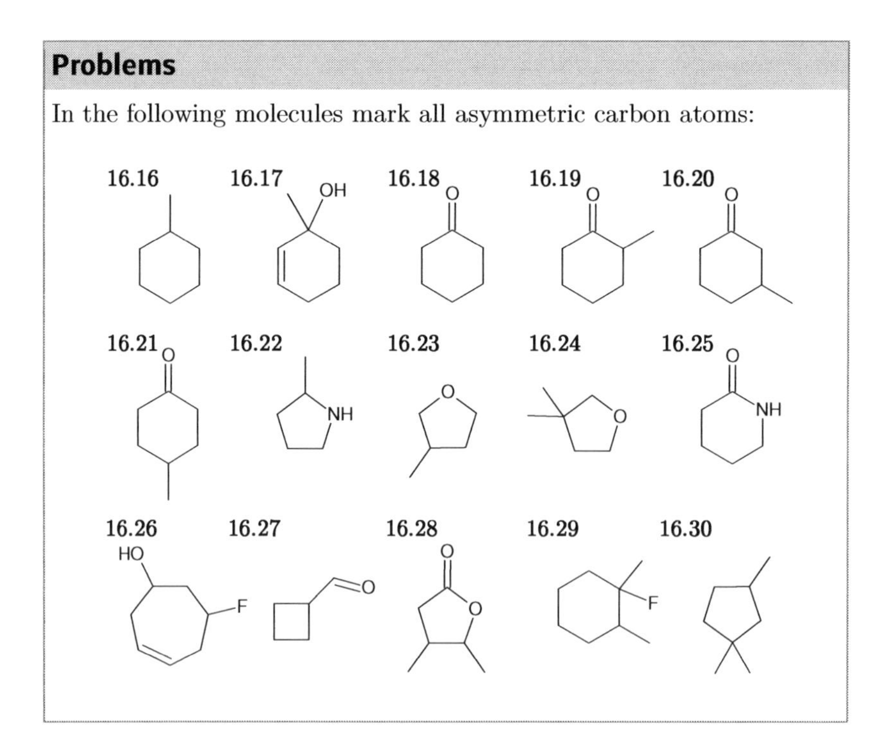

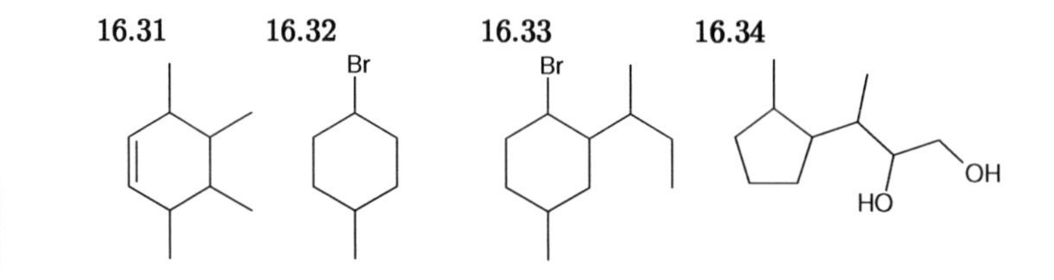

16.35 Draw a citronellol molecule and find C*.
16.36 Draw a menthol molecule and find C*s.

16.37-16.44 There are 8 different fluoroalkanes with $C_5H_{11}F$ formula. Draw all of them and find every C*.

17 *R*/*S* configurations of asymmetric C atoms

Some double bonds have substituents, which we can arrange in two different configurations. Asymmetric carbon atoms are similar to C=C bonds in this regard.

Four substituents of the C atom can be arranged in two different ways around the C** – there are two **configurations** of every C* possible (*R* and *S*), and it is a source of *R*/*S* stereoisomerism. These configurations must be and are fixed and inconvertible. How is all that possible?

For sure, no one can expect this feature of molecules, when we confine ourselves to consider molecular structures in two dimensions only. Nevertheless, in 3D, it is still hard to comprehend, even though we already have some experience in the matter. Can we really arrange four things around a little ball, C* atom, in two distinctive ways? Yes, and I am rushing with a bunch of new figures to prove it.

Look at the 3D model of 1-chloro-1-fluoroethane. There is a C* with Cl, F, CH_3 and H as substituents. I have rotated it in space:

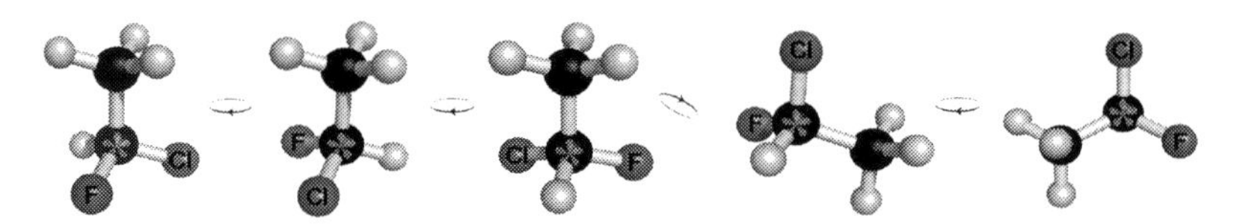

Fig. 17.1 One of two stereoisomers of CH_3CHFCl.

The drawings below depict another 1-chloro-1-fluoroethane. Are any of them identical with any one from above? No.

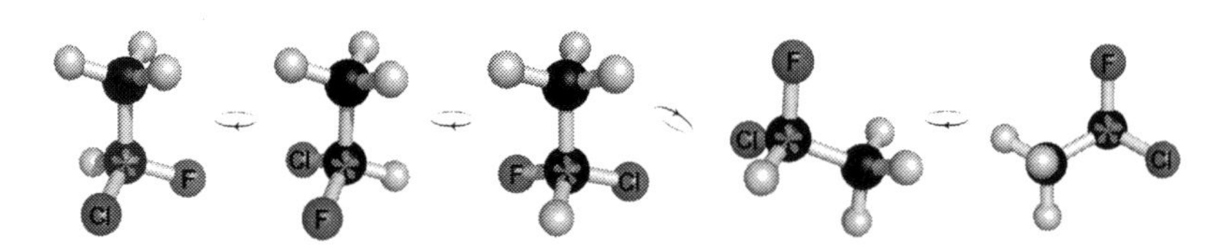

Fig. 17.2 Second stereoisomer of CH_3CHFCl.

You can rotate both models, flip and flop them endlessly. You can even compare various conformations by rotating groups along the C–C bond, but this stratagem will turn out to be pointless too. In fact, you will never find a pair of clones. Upper and lower models are truly *two different molecules.*

They differ in spatial arrangement of four substituents around C*; that is in the configuration of C*. First one will be denoted with *R* stereodescriptor, while the other *S*.

In any molecule any C* *always* is like that, and gives rise to *R*/*S* stereoisomerism. There are no exceptions (recall that such claim does not hold for C=C bonds – some of them do not have two different *E*/*Z* configurations).

How to assign *R* or *S* stereodescriptors to the configuration of C*?

We will need to rank substituents around C* according to their *priority*. We do it just like in Chapter 14, where *E* and *Z* stereodescriptors were in the foreground. Recall that "game" of fighting atoms.

1. Look at the molecule from such a direction, that the substituent with the lowest priority (usually H atom) is behind.
2. Now mark three substituents in your field of vision, with their priorities: 1st, 2nd, 3rd.
3. Assign the proper stereodescriptor:
   - *R*, when priorities increase clockwise, that is to the right (from Latin *rectus*, what means "right"),
   - *S*, when priorities increase anticlockwise, that is to the left (from Latin *sinister*, what means "left").

In 1-chloro-1-fluoroethanes, we rank substituents in order of increasing priority: Cl > F > CH_3 > H. Then we look from such a direction that H is behind. And these are our views:

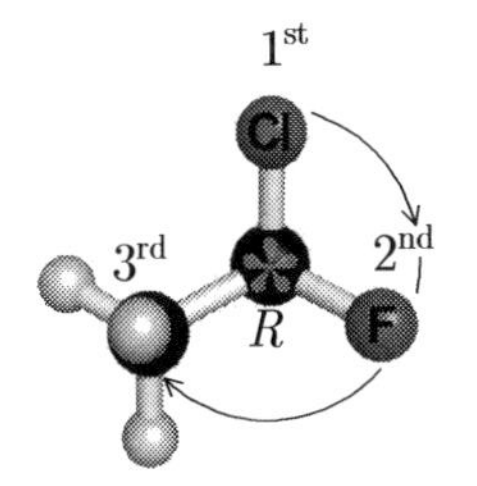

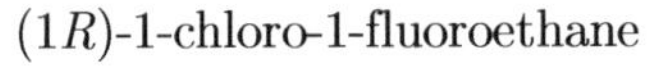
(1*R*)-1-chloro-1-fluoroethane

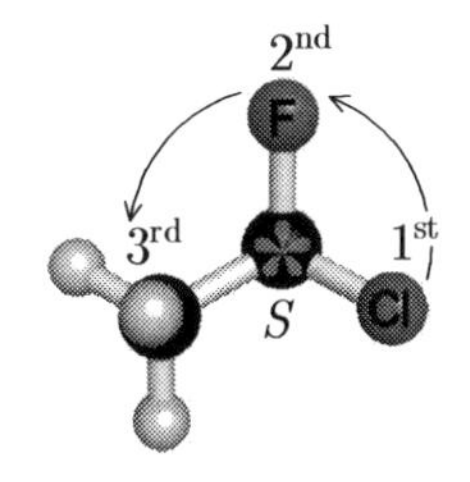

(1*S*)-1-chloro-1-fluoroethane

Fig. 17.3 Assignment of *R* and *S* stereodescriptors to two stereoisomeric 1-chloro-1-fluoroethanes.

As you see, it is not at all difficult. I hope the notion that C* can have substituents in one of two different and permanent 3D arrangements, becomes less counterintuitive now.

Anyway, such a phenomenon occurs in our daily lives too. We just do not pay attention. Look at my lunch – I had a pork chop in a coating, potato purée sprinkled with dill, and some young cabbage, which has been patiently stewed with allspice and bay leaf. There are two ways of serving the meal – two different configurations, which are fixed, and inconvertible (assuming you must not turn the plate upside down):

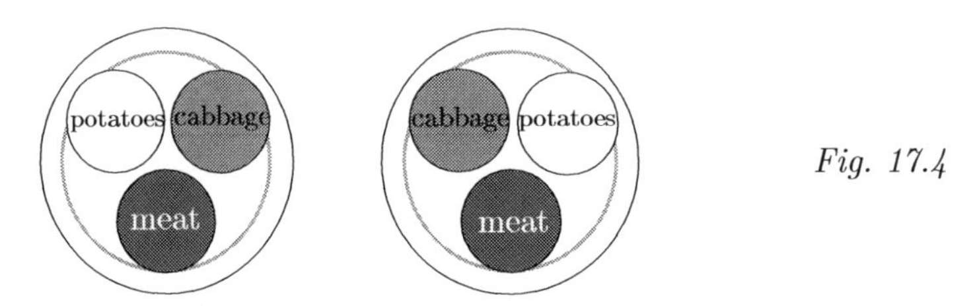

Fig. 17.4

We could even assign *R* and *S* stereodescriptors, if we had some rules of priority. Let me propose the order: 1st meat, 2nd cabbage, 3rd potatoes (according to the subjective tastiness). With this assumption left plate has *S* configuration, while the right one is *R*.

Interestingly, all chefs and waiters *know* there is a difference between both plates. The meal should be served only in one manner – meat always in front of the guest, potato at eleven o'clock, while vegetables at two o'clock (*S* configuration).

R and *S* stereoisomers on structure drawings

A configuration of a C* atom is clear on 3D models. But what about 2D structure drawings? In order not to lose that information, we use wedge and dash bonds.

This is how it works: we draw two of four single bonds as normal lines, which lie on the surface of paper. Then, one bond must go up (a wedge), and one must point down (a dashed bond). Watch carefully how I have translated 3D models of (1*R*)-1-chloro-1-fluoroethane into simpler drawings with wedge and dash bonds (Fig. 17.5).

Note that I obtained the last drawing by rotating the molecule until H hid behind C*. Formally, it is a Newman projection along the C*–H bond. In practice, it is the most comfortable "point of view" to determine the configuration.

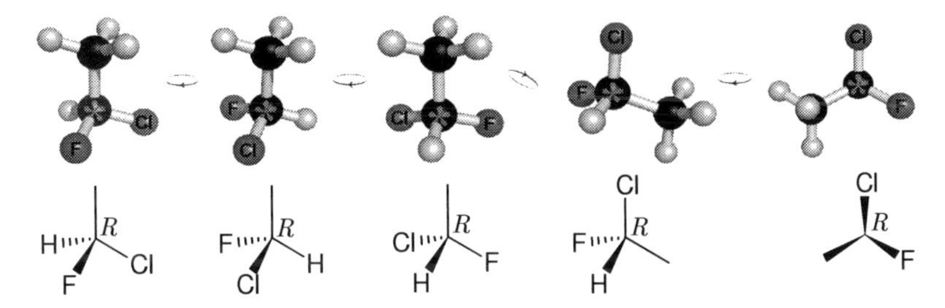

Fig. 17.5 Structure drawings of (1*R*)-1-chloro-1-fluoroethane.

To assign a stereodescriptor you can pick up whichever of the drawings. The only difficulty is that you must put your "eye" in correct position versus the molecule (recall the first step: "look at the molecule from such a direction, that the substituent with the lowest priority is behind"). Then, you have to determine how the molecule looks like from there. Consider the example, which exploits yet another structure drawing of (1R)-1-chloro-1-fluoroethane (absent on the Fig. 17.5):

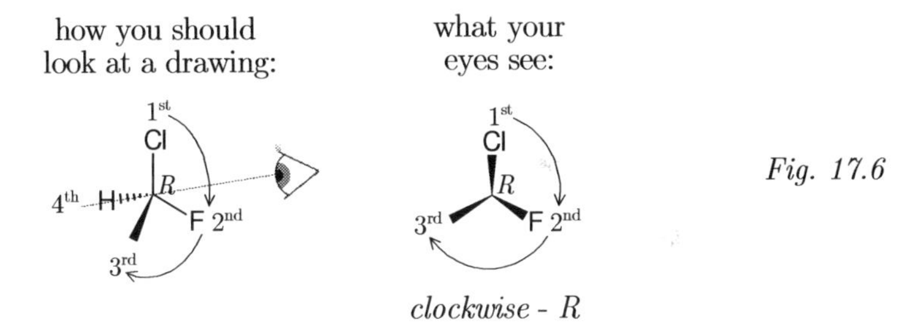

Fig. 17.6

Easy as that. Now, consider translation of 3D models of (1*S*)-1-chloro-1-fluoroethane into structure drawings:

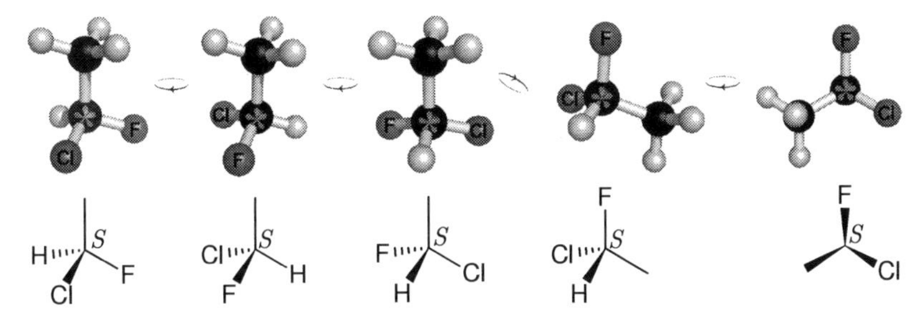

Fig. 17.7 Structure drawings of (1*S*)-1-chloro-1-fluoroethane.

Let us once again check the configuration. This time I will pick up the fourth drawing, because it seems to be the most "inaccessible." I just want to demonstrate, that the method always works:

how you should look at a drawing:

focus on what your eyes really see:

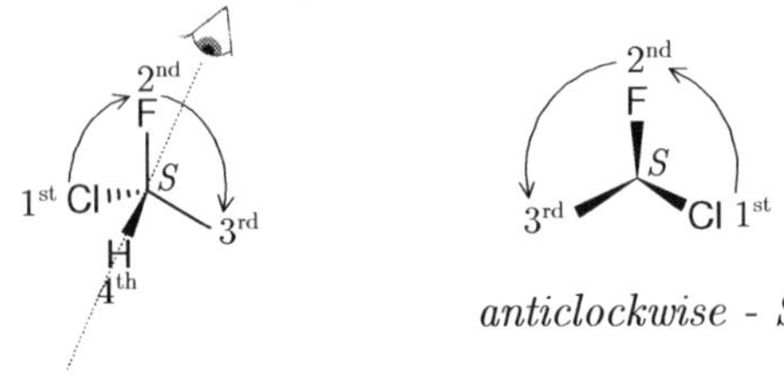

anticlockwise - S

Fig. 17.8

Problems

Assign *R* or *S* stereodescriptors to C*s in the following molecules:

17.1 **17.2** 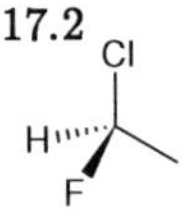 **17.3** 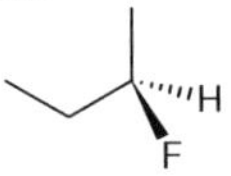 **17.4**

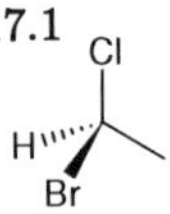

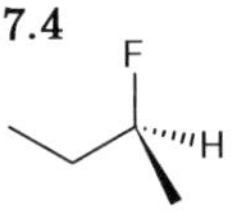

17.5 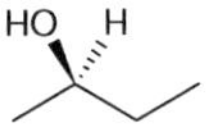 **17.6** 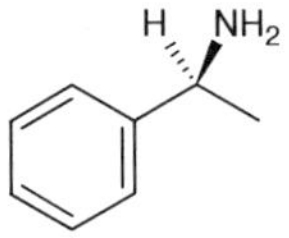 **17.7** 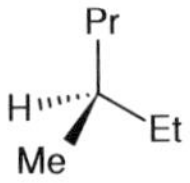 **17.8**

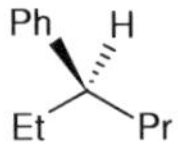

17.9 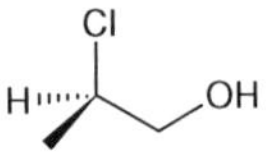 **17.10** 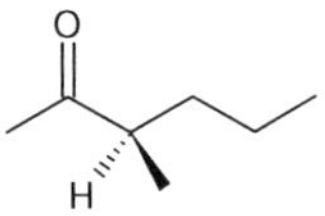 **17.11** 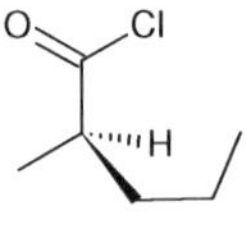 **17.12**

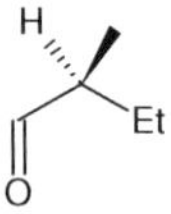

All drawings above have Hs in the back. It is quite comfortable. However, it will not always be the case – why should it? Assign *R* or *S* in structures, with happen to be a bit meticulous:

17.13 **17.14** 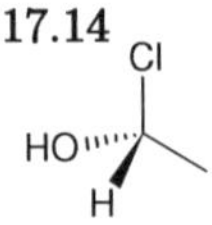 **17.15** 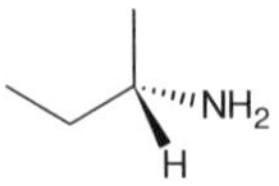 **17.16**

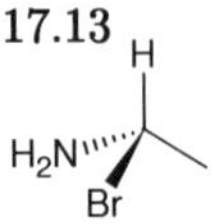

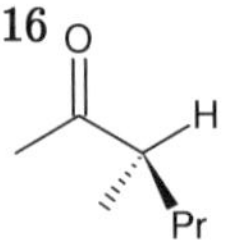

We can draw *R* and *S* stereoisomers in many ways. Please, spend some time to convince yourself that all of the following 8 drawings are 2-chloropropan-1-ols. Assign stereodescriptors to find drawings of (2*R*)- and (2*S*)-2-chloropropan-1-ols:

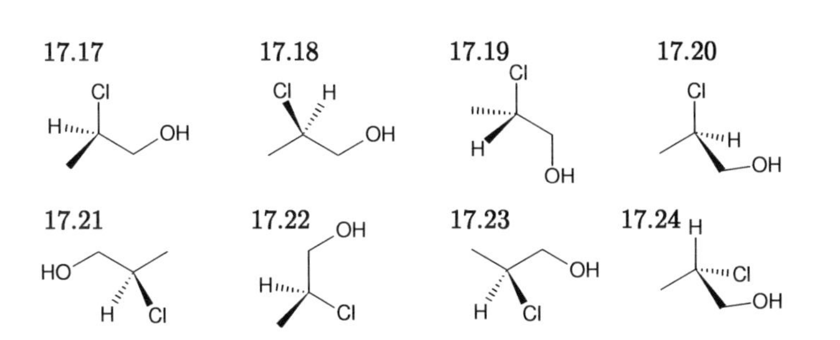

Structures below are examples of molecules with no H atom on C*. Assign *R* and *S* configurations anyway:

17.25 Cl Et Br
17.26 Cl Br NH_2 F
17.27 Pr Ph Et Me
17.28 OH
17.29 HO NH_2 OH
17.30 NH_2
17.31 H_2N F
17.32 O

Write full IUPAC names of the following molecules:

17.33 name molecule **17.31**
17.34 name molecule **17.29**
17.35 name molecule **17.28**
17.36 name molecule **17.16**
17.37 name molecule **17.13**
17.38 name molecule **17.12**
17.39 name molecule **17.11**
17.40 name molecule **17.7**
17.41 name molecule **17.6**

How to invert the configuration on the drawing?

Let us say you have a drawing of *R* stereoisomer of some molecule, and need its *S* counterpart. Nothing simpler. In order to *invert the configuration swap any two substituents*:

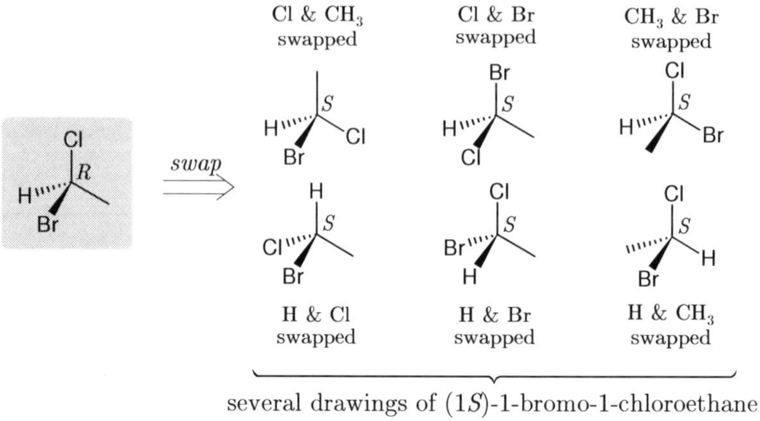

Cl & CH_3 swapped — Cl & Br swapped — CH_3 & Br swapped

swap

H & Cl swapped — H & Br swapped — H & CH_3 swapped

several drawings of (1*S*)-1-bromo-1-chloroethane

Fig. 17.9

Get rid of H atoms

We do not draw Hs attached to Cs in professional structure drawings, however, so far in this chapter, I am still dong it. There is no reason for that and we can get rid of H atoms bonded to C*s.

We remove H together with its wedge or dashed bond, and arrange remaining three, so that they fit with the honeycomb. Take a look at three examples below. I have copied different drawings of (2*R*)-2-chloropropan-1-ol from problems **17.17**, **17.18** and **17.20**:

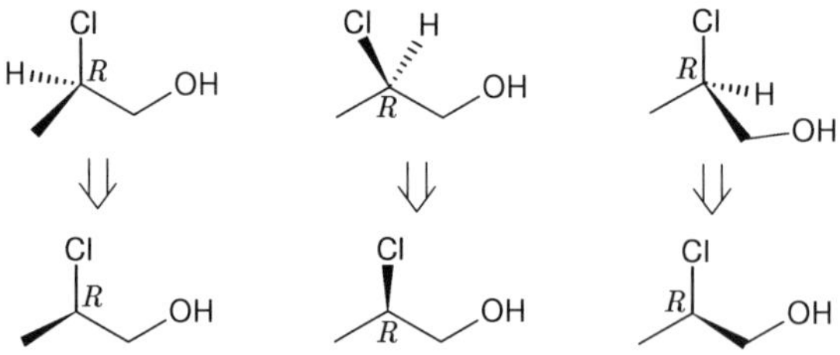

Fig. 17.10

Now, things appear clearer. As you see, the position of the wedge has no influence on the configuration. Whenever you want, *you can invert configuration* by swapping any of substituents, or by *changing the wedge bond into the dashed one*:

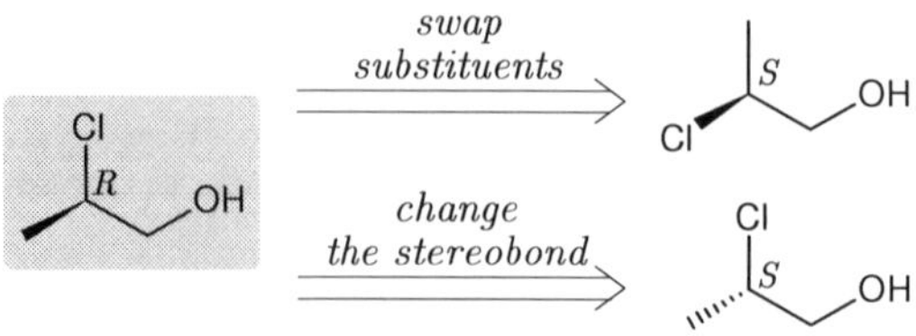

swap substituents

change the stereobond

Fig. 17.11

An additional advantage of the simplified drawing is that determination of an unknown configuration is faster. On the *drawing with a wedge, H atom must be on the back.* Therefore, we already look at the molecule from the right direction:

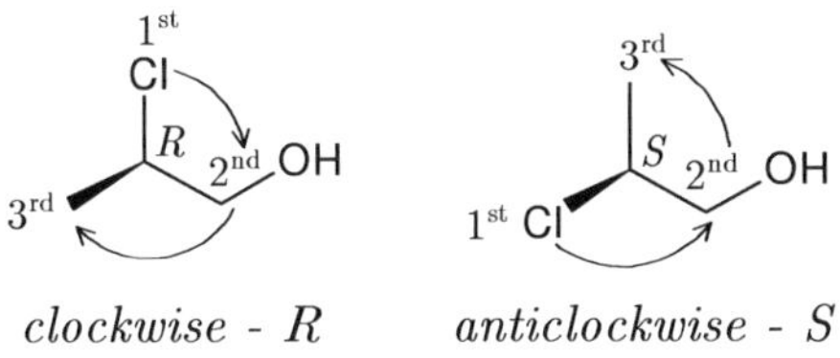

1st Cl R 2nd OH 3rd
clockwise - R

3rd S 2nd OH 1st Cl
anticlockwise - S

Fig. 17.12

On the other hand, the *drawing with a dashed bond must have H atom in front.* It bonds to C* via a hidden wedge. Therefore, you are looking at the molecule from the *wrong direction.* But be smart. If there are only two possible answers, knowing the wrong one tells you the truth, anyway:

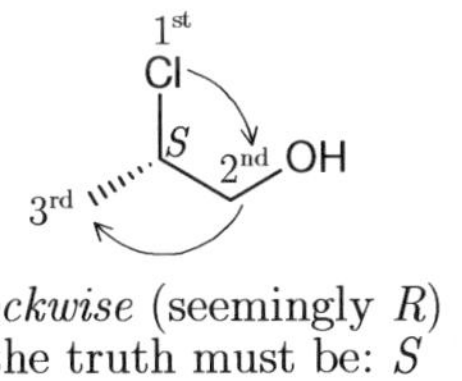

1st Cl S 2nd OH 3rd
clockwise (seemingly *R*)
the truth must be: *S*

Fig. 17.13

The truth must be *S*, because we were looking at the molecule from the wrong direction, so that the direct answer, *R*, was also wrong.

This method makes assignment of stereodescriptors to asymmetric carbon atoms faster, because we really do not need to tinker in imagination. Whenever you have a drawing with H explicitly shown, you can simply remove it, and you end with simplified structure, which is easier to handle.

Problems

Determine *R* and *S* configuration in the following molecules:

17.42 OH

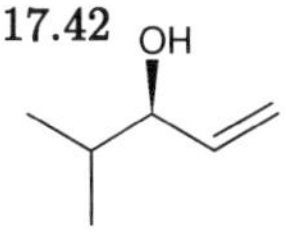

17.43

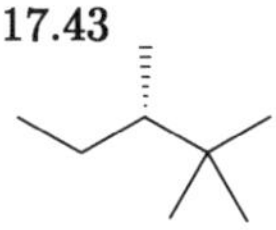

17.44 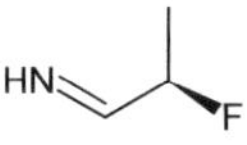 HN F

17.45 N

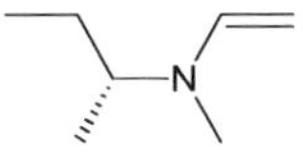

17.46 Cl O

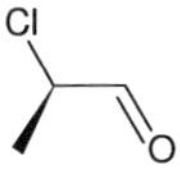

17.47 NH_2 COOH

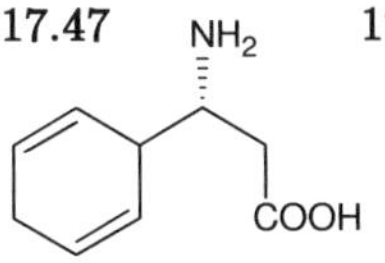

17.48 F

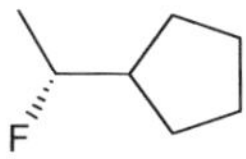

17.49 OH Et O

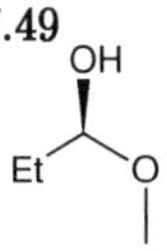

17.50 Citronellol has one C*. It means that, in fact, there are two stereoisomeric citronellols. Draw and name them.

17.51 Look at the 3D model of citronellol from the first chapter of the book (Fig. 1.2). Can you tell which stereoisomer is that?

What if a molecule has more than one C*?

R and *S* descriptors must be determined for each of C*s. Obviously, the total number of possible stereoisomers increases. For instance, structure with two C* atoms may exist in as much as four stereoisomeric versions: *R*,*R*; *R*,*S*; *S*,*R*, and *S*,*S*.

Problems

Assign stereodescriptors to the molecule and name it:

17.52

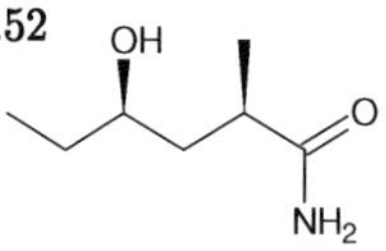

OH

O

NH_2

17.53-17.55 Draw all other possible stereoisomers of **17.52** and assign *R*/*S* stereodescriptors.

In the following aldehyde, determine configurations of a C=C bond and C* atoms, and write its IUPAC name:

17.56

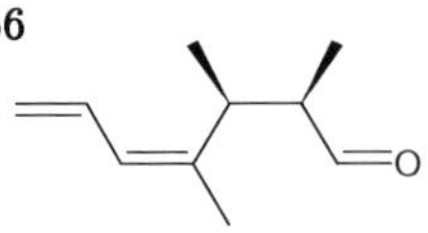

O

17.57-17.63 Draw all other possible stereoisomers of **17.56** and assign stereodescriptors.

17.64 Menthol is an example of a molecule with 3 C*. What is the total number of stereoisomers? Count all possibilities.

Try something harder. Following molecules have C* atoms as part of the ring. Assign *R*/*S* stereodescriptors:

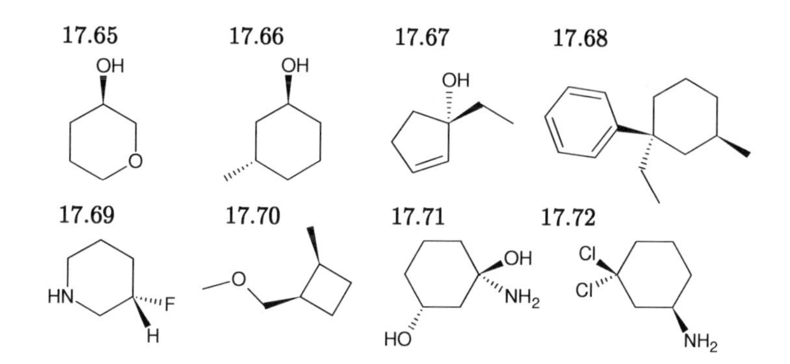

17.73 In some of the molecules above, *cis* and *trans* descriptors are applicable. Find them and describe.

17.74-77 *Cis*/*trans* system is redundant, because for many cyclic molecules both *cis* and *trans* descriptors describe two – not one – stereoisomers. Prove it: draw all R/S stereoisomers of **17.66** molecule, and check which are *cis*, and which are *trans*.

18 Enantiomers, diastereoisomers and chirality

Every R isomer and its S counterpart *are mirror images of each other.* To show you that, I drew (S)-citronellol, put a mirror on its right side, and then drew the mirror image:

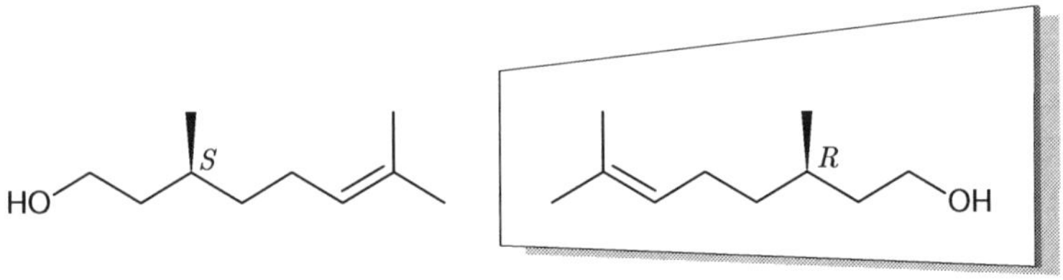

Fig. 18.1 (S)-citronellol and its mirror image, which turns out to be (R)-citronellol.

You do not have to trust me – assign stereodescriptors at your own to convince yourself that the left drawing is S, while its mirror image is R stereoisomer. Otherwise, we can also flop the right drawing, to make the thing clear. Note that the wedge bond goes beneath the surface of paper, so in the flopped drawing it becomes a dashed bond:

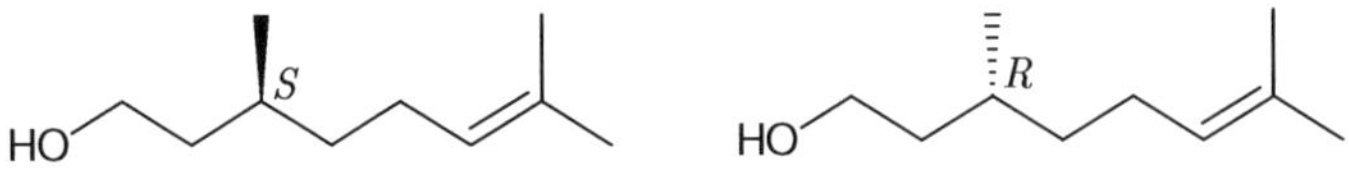

Fig. 18.2

Now, recall my lunch from the previous chapter – the plate with meat, potatoes and cabbage. They both differ in configurations, and both are mirror images of each other. Or take a look at your hands. Left hand differs from right hand, because of the configuration (arrangement of every cell around the center of a palm), and once again, they are mirror images of each other.

Objects which exist in two "versions", which are nonidentical mirror images of each other are chiral

The problem with the notion of **chirality** results from the fact that people tend to believe in identicalness of things and their mirror images. This is true only for some objects, which we call **achiral**: sphere, perfect scissors, a pencil, a cup of coffee, or a perfectly symmetrical bottle of wine. Put one in front of a mirror, and you will not be able to see any difference.

However, most real things are **chiral**: my lunch, our hands, ears, gloves, shoes, are some of most obvious examples. They all are not identical with their mirror images. Moreover, the face you see in the mirror is not exactly your face too...

We can describe every physical object as chiral or achiral, and molecules are no exception:

chiral molecule	**achiral molecule**
is a paired object	is a lonely object
whenever it looks into the mirror it sees its companion	whenever it looks into the mirror it sees itself
likes the fact that it and its companion are so similar, yet totally differ in configurations	does not even want to know why there is a fuss about these "configurations," or something
likes the fact that chemists curiously call it and its companion: **enantiomeric pair**	hates the fact that there are no fancy words about it, for students to learn

To sum it up - a molecule is chiral whenever it has a pair, a companion, which is its mirror image – a similar molecule, though totally differing in the configuration. We call these two an enantiomeric pair or, simply, **enantiomers**.

Enantiomeric pairs and diastereoisomeric pairs

Every pair of *R* and *S* stereoisomers is an enantiomeric pair. When *R* looks into the mirror, it sees *S*. When *S* looks into the mirror it sees *R*. The chirality of these molecules arises from the presence of an asymmetric carbon atom and its different configurations.

And what about molecules which have more than one C*? When some *R*,*R* stereoisomer looks into the mirror it sees *S*,*S* – enantiomeric molecules differ in configurations of every C*. Therefore, *R*,*S* stereoisomer, looking into the mirror, sees *S*,*R* as a companion. *R*,*S* and *S*,*R* are also a pair of enantiomers. And both must be chiral molecules too.

However, while C* is a source of chirality, the double C=C bond, which can also have two different configurations, is *not*. When some *Z* alkene looks into the mirror, it sees itself... Sadly, the same happens for *E*. None of them has an enantiomeric companion. They are unpaired, lonely objects, and achiral molecules. Proof:

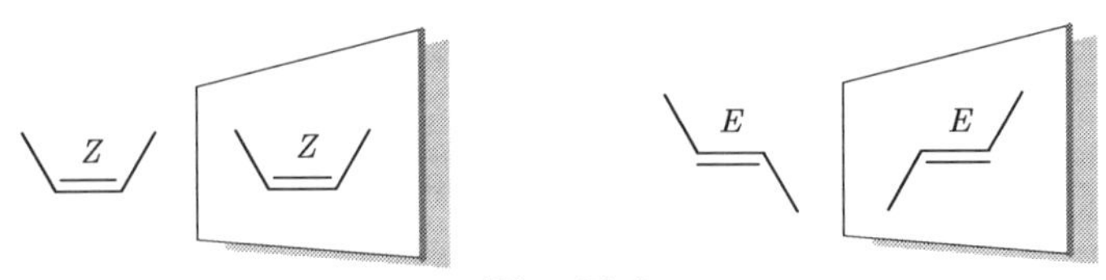

Fig. 18.3

Chemists invented a name for any *pair of stereoisomers, which are not a pair of enantiomers*: the **diastereoisomeric pair**.

While we can imagine an enantiomeric pair as a close relationship between molecules, which are mirror images of each other, the diastereoisomeric pair is a rather emotionless association. To be honest: stereoisomers in diastereoisomeric relations do not really care about each other.

E and *Z* isomers are always diastereoisomers. On the other hand, *R*,*R* stereoisomer has its enantiomeric companion in the person of *S*,*S*, and does not care about, let's say, *R*,*S*, because *R*,*R* and *R*,*S* are only a diastereoisomeric pair. The same when you put *R*,*R* next to *S*,*R*; or *S*,*R* next to *S*,*S*; or *S*,*S* next to *R*,*S*...

Analysis of relationships between stereoisomers

With one and only one C* per molecule, there are only two stereoisomers: *R* and *S*. They are always an enantiomeric pair. Period. In the collection of stereoisomeric molecules containing two or more C*s per molecule, we will find some enantiomeric pairs, and much more diastereoisomeric relations. Consider the example:

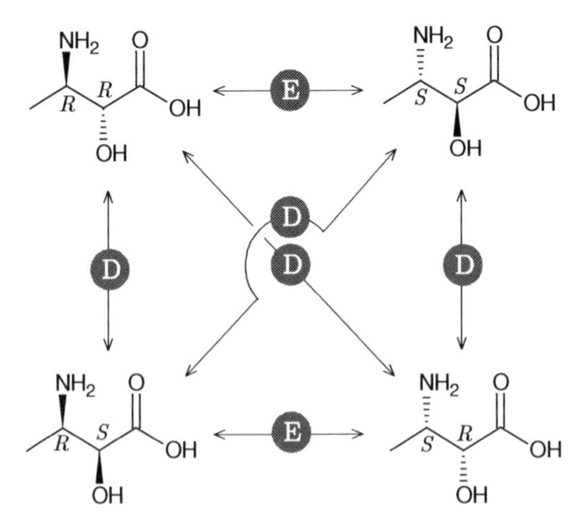

Fig. 18.4 –(**E**)– connects any two stereoisomers which are an enantiomeric pair, while –(**D**)– denotes diastereoisomeric relation.

You can always prove that two enantiomers are indeed companions "from the mirror." Let's check one of allegedly enantiomeric pairs from the Fig. 18.4: *R*,*R* and *S*,*S*. I redrew *R*,*R* stereoisomer, put a mirror on its right side, and then drew the mirror image. Indeed, it is the *S*,*S* stereoisomer from the above figure, though the drawing must be flopped to make it explicit:

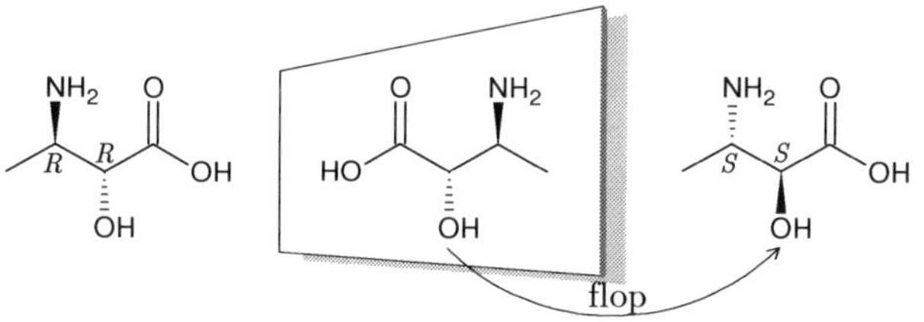

NH_2 O R R OH OH | O NH_2 HO OH | NH_2 O S S OH OH — flop

Fig. 18.5

Fig. 18.6 All new terms gathered in one diagram:

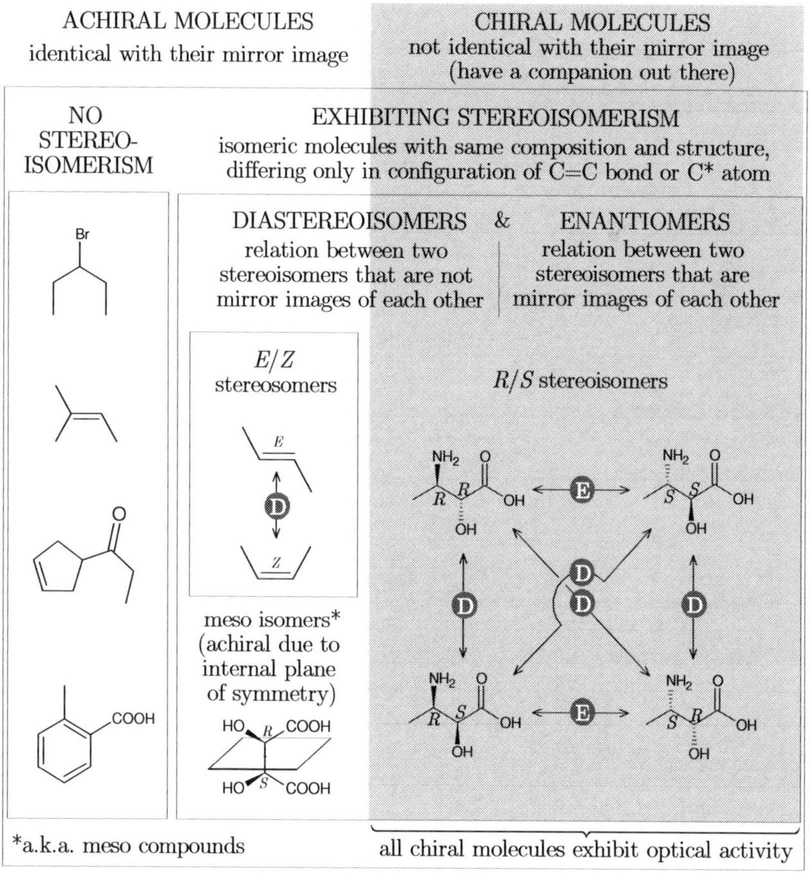

ACHIRAL MOLECULES
identical with their mirror image

CHIRAL MOLECULES
not identical with their mirror image
(have a companion out there)

NO STEREO-ISOMERISM

EXHIBITING STEREOISOMERISM
isomeric molecules with same composition and structure, differing only in configuration of C=C bond or C* atom

DIASTEREOISOMERS &
relation between two stereoisomers that are not mirror images of each other

ENANTIOMERS
relation between two stereoisomers that are mirror images of each other

E/*Z* stereosomers

E, D, Z

meso isomers*
(achiral due to internal plane of symmetry)

HO R COOH
HO S COOH

R/*S* stereoisomers

Br

COOH

*a.k.a. meso compounds

all chiral molecules exhibit optical activity

Optical activity and the racemate

Every chiral molecule exhibits optical activity. **Optical activity** is a unique physical property, which relies on the ability of a solution of chiral molecules to *rotate the plane of a polarized light.* We are not going into theoretical details of the process (fortunately). I will rather outline the stuff you *need* to know.

Chemists use an apparatus called **polarimeter** to test optical activity. A beam of polarized light points at the solution, and if there are chiral molecules inside, the plane of light rotates, as it goes through. We measure the angle of rotation, when the beam gets out.

Interestingly, when one particular chiral stereoisomer rotates the plane of the light by, let's say $+30°$, its enantiomeric companion, tested in subsequent experiment, will rotate the light by exactly opposite value of $-30°$. Here is a link to concise video about optical activity if you want to delve into some more details:

http://www.youtube.com/watch?v=8alTLZJHf9k.

A mixture of two enantiomers, carefully prepared so that the ratio is 1:1, *will not* rotate the light, because the action of one molecule of the first stereoisomer perfectly cancels the action of one molecule of the second stereoisomer. The mixture, as a whole, is not optically active, although each of the components is. We call such mixtures **racemic** (or simply a **racemate**).

Some chemists like to indicate the sign of the rotation of light with (+) or (–) signs. They put it in front of the name of a molecule, instead of *R* or *S* stereodescriptors, to make our lives harder.

You should never confuse + and – with *R* and *S* stereodescriptors. Stereodescriptors come from theoretical assignment, while + and – are from the physical experiment. There is no direct connection between these two distinctive concepts. And although you can assess *R* or *S* configurations by inspecting the drawing, you cannot predict how the molecule rotates the plane of light until you do an experiment.

Racemic mixture can be denoted as +/– or ±, and this symbol is quite useful. If you have a bottle of racemic mixture of citronellol it is convenient to write (±)-citronellol on its label, instead of writing "a 1:1 mixture of (*R*)-citronellol and (*S*)-citronellol."

Problems

18.1 The molecule with *n* stereocenters (C* atoms or C=C bonds) a maximum of 2^n stereoisomeric versions possible. Use this formula to count how many stereoisomers a menthol molecule has.

18.2 Draw all stereoisomeric menthols. Point out all enantiomeric pairs. Answer the question about each of the structures: is it chiral? (yes or no), is it optically active? (yes or no).

18.3 Interestingly only one of all stereoisomeric menthols is a natural molecule occurring in the peppermint. This is exactly the one shown in the 3D picture in Chapter 1. Assign R/S stereodescriptors to each of C*s in this natural menthol.

19 *R/S* stereoisomerism in nature and medicine

Molecules in Nature are often so structurally intricate, that finding asymmetric C atoms within them is never a surprise. In fact, most have at least one C*, and exhibit *R*/*S* stereoisomerism.

For example, **testosterone**, the male sex hormone, has a backbone made of 19 C atoms. As many as six of them are asymmetric, so there are $2^6 = 64$ different stereoisomers. Remarkably, from all these 64 possible versions, our organisms synthesize only one stereoisomer:

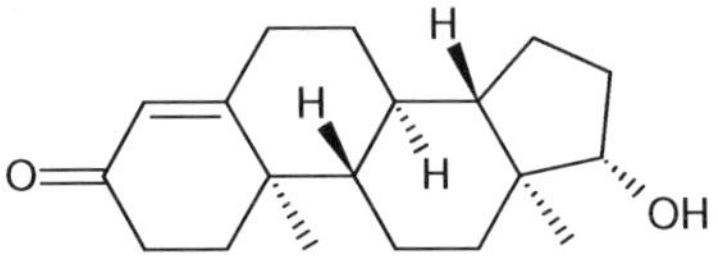

Fig. 19.1

(8*R*,9*S*,10*R*,13*S*,14*S*,17*S*)-testosterone

This molecule is the only one, super pure testosterone, which floats in your bloodstream. The feature is universal – all famous biomolecules like DNAs, RNAs, or proteins *are always characterized by stereochemical purity.* The consequence of this fact is profound, because:

Pure stereoisomer of one molecule differentiates pure stereoisomers of other molecules

Example: think of our hands as two stereoisomers of a molecule X, and about gloves as two stereoisomers of a molecule Y. These objects can differentiate between each other. That is why the left glove fits your left hand perfectly, while the same cannot be said, when you put it on your right hand.

Since biomolecules inside our bodies are stereochemically pure, we can expect them to response and interact variously with different stereoisomeric versions of external molecules. Just like our hands response differently to two isomeric gloves.

Glutamic acid is a good example. This substance is frequently added to food as flavor enhancer. It is responsible for one of five basic tastes – *umami.* Read up on it in the web it if you thought, that sweet, salty, sour and bitter were all basic tastes out there.

Interestingly, only *S* stereoisomer of glutamic acid has a flavor enhancing ability, while *R* is useless. Taste receptors on our tongues are made of stereochemically pure biomolecules, which recognize the difference.

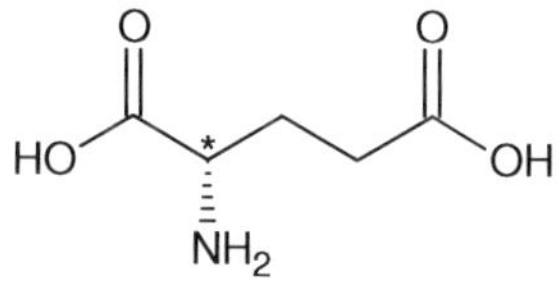

(*S*)-glutamic acid

Fig. 19.2

This diversification of responses to different stereoisomers happens to be a source of danger. In the fifties, chemists from Germany produced a new sedative drug for pregnant women. They called this infamous molecule **thalidomide** and marketed as a wonder drug for insomnia, sickness and headaches.

Pharmacologic tests showed that thalidomide worked well, and they introduced it to the market in 1957. Unfortunately, by the year 1961, when it was withdrawn, mothers who used thalidomide during their pregnancy gave birth to about 12 thousand children with severe limb deformities. Today these people are in their fifties. The pharmaceutical company issued the apology in 2012.

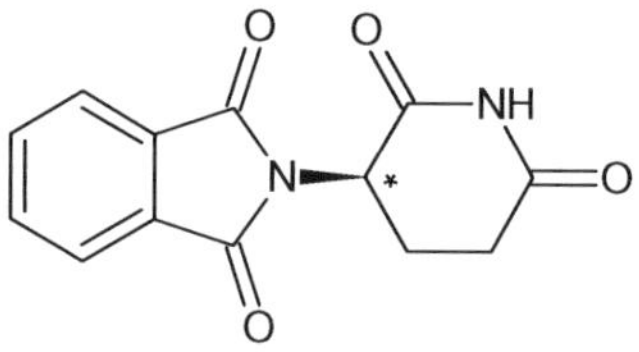

(*R*)-thalidomide

Fig. 19.3

Thalidomide disaster was associated with stereochemical nature of the molecule – there is one C*, so there are two thalidomides... In the process of chemical synthesis both form at once in 1:1 ratio. Such a racemic drug was applied to patients. According to later research, only *R* isomer was a beneficial medicament, while *S* gave rise to tragedy. Within the body, actions of (*S*)- and (*R*)-thalidomide were different. *S* selectively interacted with DNA of fetuses, triggering its malfunction.

One may think: okay, let's remove *S* stereoisomer from the drug, and, eventually, we have a great product to sell. It can be done but... will not solve the problem.

Investigations showed that when super pure (*R*)-thalidomide is introduced to the body, it becomes transformed into 1:1 mixture of *R* and *S*. Chemists call such process **racemization**.

How could that happen, if the configuration of C* is fixed and inconvertible? Racemization is a *chemical reaction* in which chemical bonds around C* are constantly broken, and reformed.

Meanwhile, substituents are free to change their spatial arrangement. In consequence, after bonds reformation, C* atoms in reformed molecules may have starting *or opposite* configuration.

The process may occur to stereoisomeric molecule, and it depends on its chemical nature, as well as the environment in which we place it. Fatefully, internal environment of our bodies turned out to be perfect for racemization of thalidomide.

Asymmetric synthesis

The case of thalidomide demonstrates the difference between how Nature synthesizes molecules, and how people do it.

Nature produces stereochemically pure substances, while *reactions in chemical laboratories usually lead to racemic mixtures.* It is not always like that, but it is more the rule than the exception.

Fortunately, chemists developed various methods to separate stereoisomers from each other. Simultaneously a relatively new branch of chemical science, which we call an **asymmetric synthesis**, is being developed. The purpose of asymmetric synthesis is to design new universal organic reactions and procedures, which lead to one, pure stereoisomer of the product. They ought to replace old methods, which give rise to racemates, in order that we can produce drugs perfectly suiting needs of our stereochemically pure organisms. This pursuit is just one of examples of human struggles to mimic Nature's extraordinary skills and performance.

Part V RESONANCE AND INDUCTION

20 Electrons in molecules: σ, π and lone pairs

We leave 3D space now, and go back to flat drawings of molecules to focus on electron pairs. As you know, the structure drawing shows explicitly all chemically active electrons:

Chemically active electrons: σ, π, and lone pairs

Every structure drawing is made of strokes representing σ pairs of electrons – the structural foundation of the molecule. There are additional strokes, put here or there, which are π pairs in double and triple bonds and, optionally, dots representing lone pairs (lp for short). Normally, they do not participate in formation of chemical bonds, because their atoms already have the octet. Catch the difference between these three categories:

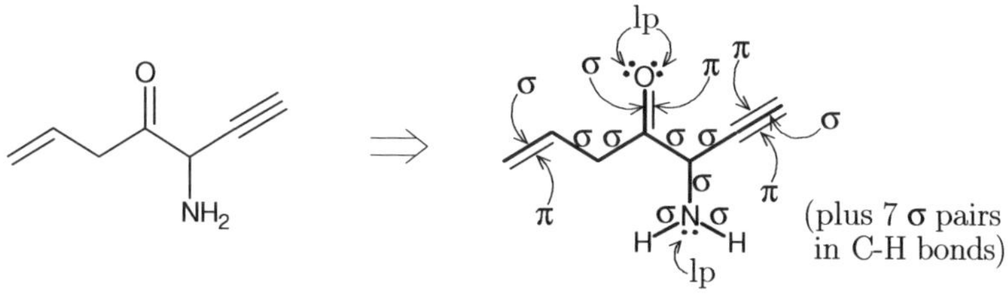

O
NH_2
⇒
lp
σ σ π π σ
σ σ σ σ
σ
π π
σ N σ
H H
lp
(plus 7 σ pairs in C-H bonds)

Fig. 20.1

As you see, σ pair is the first bonding pair in every chemical bond, while any bonding pair beyond the first one is π. Thus, every single bond is made of σ pair; double bond is made of one σ and one π bonding pair; while one σ and two π pairs constitute a triple bond.

Problems

In the following molecules mark every electron pair as σ, π, or lp:

20.1 20.2 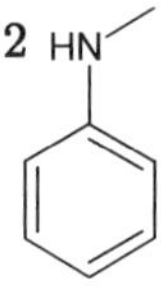 20.3

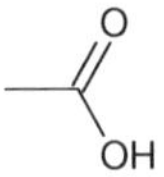

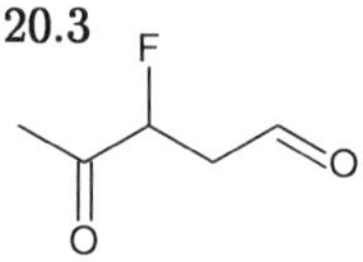

20.4  20.5

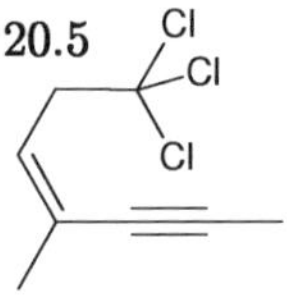

We can call the entire conglomeration of chemically active electrons of a molecule: σ, π and lp – its **electronic structure** (although, formally, it is a far-reaching simplification).

Significance of σ, π, and lone pairs

Chemical reactions are usually associated with the change of the atomic composition. However, it is not always the case, and the most general definition is that each *chemical reaction is the change of the electronic structure of the molecule.* Therefore, electron pairs must be essential for our understanding of organic chemistry. They determine chemical properties of molecules.

To delve into the electronic structure of molecules is necessary. Otherwise, you will not be able to understand organic reactions. This very part is devoted to consideration of roles, which σ, π, and lp electrons play within the molecule. Two pivotal notions we are going to familiarize with are **resonance** and **induction**:

	π pairs & lone pairs	σ pairs
phenomenon	delocalization of π/lp electrons & *resonance*	polarization of σ bonds & *induction*
meaning	In some molecules π pairs & lone pairs are delocalized, but structure drawing fails to show that; anyway, we must be able to find and recognize delocalized electrons, and use resonance method to show how it influences the molecule.	In all molecules σ bonding pairs are polarized giving rise to partial charges on atoms, though structure drawing fails to show that; anyway, we must be able to predict the distribution of charges, and how this induction affects the molecule.

21 Delocalization of π electrons and lone pairs

All drawings depict π pairs and lone pairs as *localized.* We can say, "this π pair is localized exactly between this and that atom," or "this lone pair is localized exactly on that N." Seems justified, because this is exactly what the drawing shows. However, such assessments are not always correct.

There are molecules in which π or lp electrons are **delocalized**. It is not a rare phenomenon, and there were a lot of such molecules in the book already. Delocalization means that particular π and/or lp electrons *are not located exclusively at places indicated by the drawing, because they spread over some larger area.* For the purposes of this book, I will call this area an "area of delocalization."

How to find delocalized electrons on the structure drawing

Chemical drawings are imperfect, but the situation is not hopeless. Although delocalization is not show explicitly, there is a simple trick helping us to distinguish delocalized electrons from all others. Straight to the point: you can always recognize delocalized π and lone pairs, by looking at the drawing and searching for...

...any **π pairs** of electrons, or any **lone pairs**, or any ⊕ **formal charges***...	...that are separated from each other by **one σ bond**.

(*do not worry about mysterious "⊕ formal charge" for now)

Let us try this in practice. Look at the amine below. It is one of many molecules with delocalized electrons inside. We can easily discern π pair separated from the lone pair by exactly one σ bond:

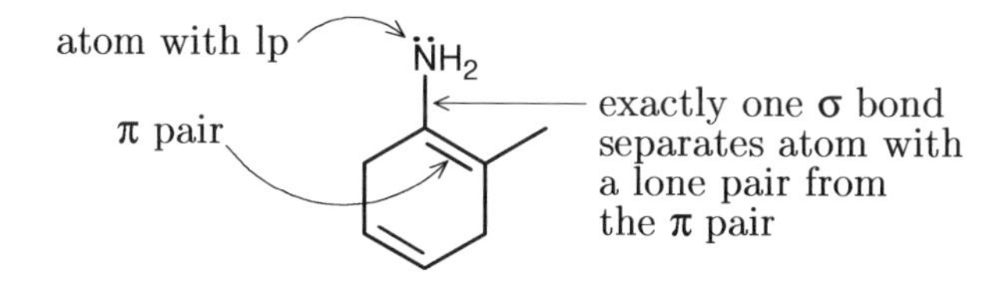

Fig. 21.1

Conclusion: lp electrons and the π pair are delocalized together.

Hint 1: lone pairs on halogens never count. Formally they are delocalized, but the phenomenon is negligible in this case.

Hint 2: two or more π pairs can be delocalized without the participation of lone pairs.

The area of delocalization

Let us go back to the example. The expression "lone pair and π pair are delocalized together" means that they are not exclusively in positions indicated by the drawing. In reality, they "mix" and spread over the area of delocalization, which starts on the N atom and ends on the C2 atom:

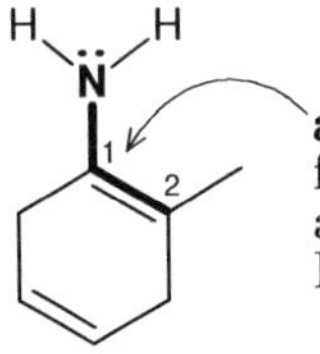

H H N 1 2

area of delocalization:
four delocalized electrons (lp and π) are delocalized around σ bonds of N-C1-C2 fragment of the molecule

Fig. 21.2

In the following chapters, I will indicate areas of delocalization with bolded σ bonds, just like in the picture above. And it will always be like that: **area of delocalization** *starts and ends on atoms with electrons we recognize as delocalized, and contains an entire fragment of the σ skeleton between.*

Our four electrons – two π and two lp – are confusingly drawn in localized places, while in reality they are "mixed" and spread over N–C1–C2 area. By the way, notice that there is a second π pair between C4 and C5 atoms. It is a very normal, localized pair, because it does not feel any π/lp electrons in the distance of exactly one σ bond.

Sometimes we refer to such normal, localized lone pairs and π pairs as **isolated**, while these, which become delocalized, are called **conjugated** (with each other).

Problems

Look at every molecule below and determine whether it contains delocalized electrons. If yes, show which ones are delocalized and mark the delocalization area by making the appropriate fragment of the σ skeleton bold. Remember: halogens do not count. Also, do not act routinely: no one said that in delocalized "system" there must be two pairs of electrons only...

21.1 NH_2

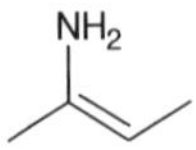

21.2 NH_2

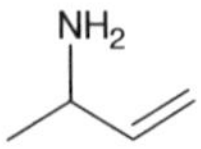

21.3 NH_2

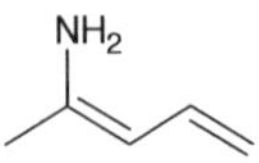

21.4 NH

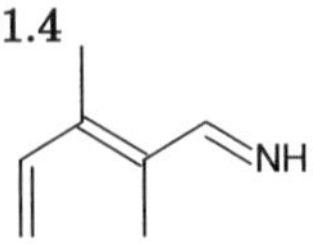

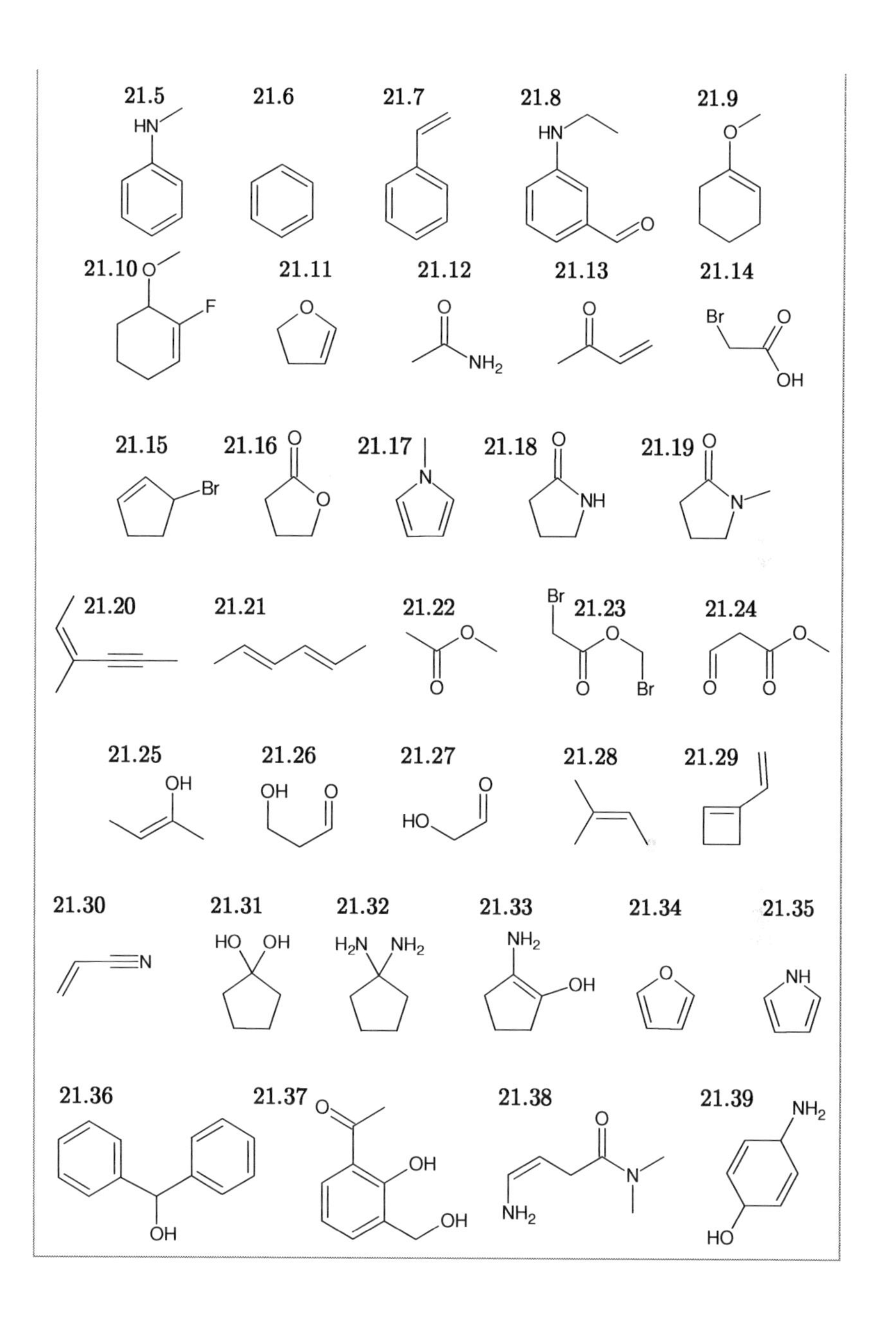
21.5
HN
21.6
21.7
21.8
HN
O
21.9
O
21.10
O
F
21.11
O
21.12
O
NH2
21.13
O
21.14
Br
O
OH
21.15
Br
21.16
O
O
21.17
N
21.18
O
NH
21.19
O
N
21.20
21.21
21.22
O
O
21.23
Br
O
O
Br
21.24
O
O
O
21.25
OH
21.26
OH
O
21.27
O
HO
21.28
21.29
21.30
N
21.31
HO
OH
21.32
H2N
NH2
21.33
NH2
OH
21.34
O
21.35
NH
21.36
OH
21.37
O
OH
OH
21.38
O
N
NH2
21.39
NH2
HO

22 How to draw resonance structures

The notion of delocalized electrons is necessary to explain specific properties of many molecules, and you will need such a skill in the future. However, there is something more – I showed you how to find delocalized electrons, and now it is time to see how to depict a delocalization.

We refer to *the method to show a delocalization* as **resonance**. Students perceive resonance as one of the most fearsome words in organic chemistry, but it is unjustified. Let me explain:

The idea behind the resonance method

Single drawing fails to show a delocalization, because we must put π and lp electrons in particular localized places. To go around this limitation we use resonance. It relies on drawing *the set of drawings*, instead of just one. It is a bit time-consuming solution, but an effective one.

We will subsequently redraw the molecule. On each drawing, we put electron pairs, which are delocalized, in different places, though only within the area of delocalization. We refer to such drawings as **contributing structures** or **resonance structures** (the latter term is more popular, and from now on, I will use it exclusively).

The set of resonance structures is closer to real nature of the molecule than the single structure drawing, because we can see that delocalized electrons are in fact here, and there, and somewhere else...

Drawing resonance structures

At first, we find delocalized electrons and determine the area of delocalization, by searching for any π pairs or lone pairs separated from each other by exactly one σ bond. Any drawing of a molecule with delocalized electrons is always one of its resonance structures. Hence, we can treat the one we start from, as the first member of the set. We will use a simple amine as an example:

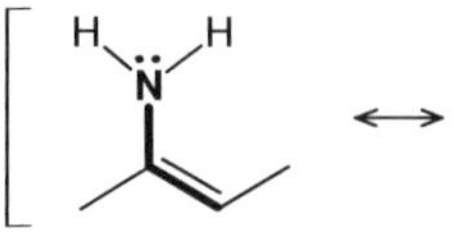

H H N ⟷

Fig. 22.1

We always put the set in square brackets, so I have drawn the opening one already. There is a **double-headed arrow** also. We draw it between every pair of resonance structures.

Now, I have two warnings for you. Remember, that in resonance of simple molecules *we move all electrons at once.* In the new resonance structure, all of them will appear with changed positions. None stays in place.

Secondly, *pushing electrons beyond the area of delocalization is strictly forbidden.* The area of delocalization is our only playground in resonance, while in the rest of the molecule nothing must change.

Okay, so how can we draw the second resonance structure of our amine? We must mutually move delocalized pairs into some new positions. So-called **curved arrows** come with help. A curved arrow grabs electron pair with its tail, while the arrow's head shows where the pair is going.

Let us start from the lone pair. It lies on the N atom, which is the terminal atom of the area of delocalization. Therefore, the only direction it can move is *toward* the center of delocalized area. We push it to the adjacent space between N and C atoms, so that it is going to become a newly born π pair in the next resonance structure:

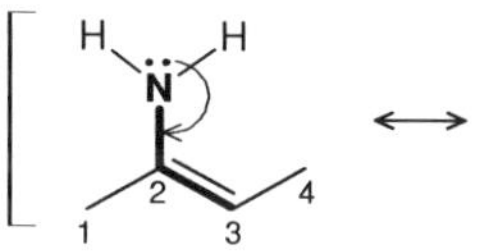

Fig. 22.2

However, be careful. As I said before, all delocalized electrons move at once, so the π pair in the C=C bond must be pushed simultaneously. It looks like *incoming lp repels π pair* of the C=C bond.

Since delocalized area ends on the C3 atom, the only place where the π pair can run away is this particular C3. Our π pair is forced to hop on it, becoming a lone pair in the new resonance structure:

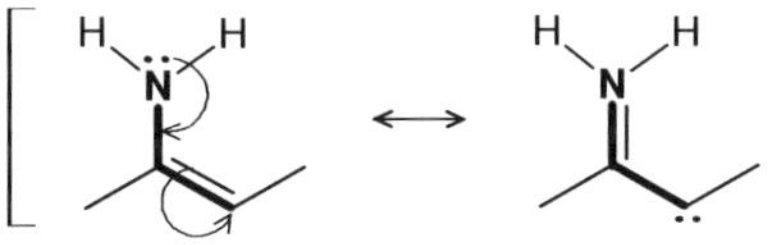

Fig. 22.3

Forgetting about both warnings will ruin your efforts: (1) never push electrons beyond delocalized area. That is why the only direction for N's lone pair was toward the center. The only thing π pair of the C=C bond could do, was to hop on the terminal C atom. (2) Move delocalized electrons simultaneously. Look at what would happen, if you decided to push lone pair from the N atom to the N–C bond, without pushing away the C=C π pair:

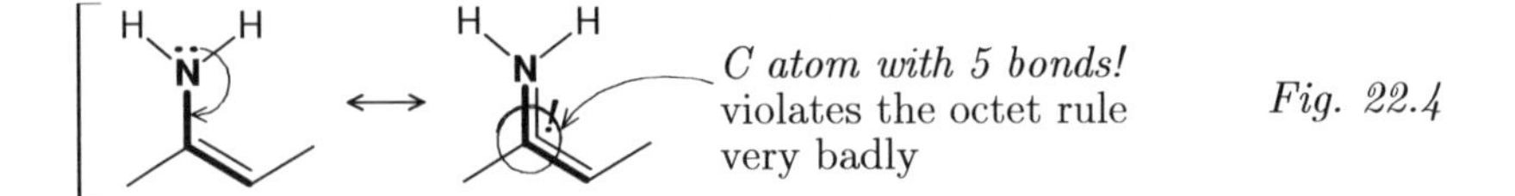

Fig. 22.4

5-valent C has 10 electrons around, and therefore violates the octet rule. Nothing like that can exist, and the drawing should be cut out from the page and burned in the hellfire. The π pair from C=C *must* run away. Moreover, because there is no other place to go, it *must* hop onto the C3 atom.

Such carbon atom with a lone pair may look a bit strange to you. We have never seen it before. Nevertheless, believe me, C atoms with lone pairs exist. The crucial fact is that such C has only three single bonds left (look at the drawing to convince yourself). It means the octet rule *is not* violated, because we have exactly 8 e⁻ around there: $2 \cdot 2$ in C–C σ bonds, 2 in undrawn C–H σ bond and 2 as the lone pair. Everything is fine.

Formal charges appear

Within the resonance structures, atoms, which experience inflow or outflow of electrons, gain so-called **formal charges**. We denote them with ⊕ and ⊖ symbols.

After an outflow of electrons, in the new resonance structure, the atom will bear a positive formal charge. We mark it with ⊕. This is exactly what happens to the N atom. It had two electrons in the lone pair *for its own*. In the second resonance structure, that pair becomes π bonding pair, and as such is being *shared* with a C atom. From the point of view of N, it is like giving away one of its electrons. Nitrogen feels an outflow of negatively charged e⁻, and is no longer neutral – gains a positive formal charge.

On the other hand, after some atom experienced inflow of electrons, it gains a negative formal charge in the new resonance structure. We mark it with ⊖. It happens to the C atom at the end of the area of delocalization. The π bonding pair, which was *shared* with the middle C2 atom, now becomes a lone pair, which *belongs exclusively* to C3. It felt the inflow of negatively charged e⁻, and now bears a negative charge:

Fig. 22.5

Note that middle C atom does not experience inflow or outflow of electrons. It has one π pair in both structures.

Remember that the **overall charge** *of all resonance structures must be the same* (overall charge is a sum of formal charges of all atoms; atoms without ⊕ or ⊖ have formal charge of 0).

The overall charge of the first resonance structure was 0, and it must remain as such. In the second structure the overall charge is 0 too, because plus and minus cancel each other. Everything is okay. It is a good practice to check the balance of charges in resonance structures you draw.

Do we need another resonance structure?

When you finish the new resonance structure, you should decide whether it is the final one. Here, the answer is yes. Why? Because there are no more places in the area of delocalization to push electrons further. The lone pair on C is already at the very end of the area.

Therefore, we can draw a square bracket on the right, stretch a little and go get a cup of coffee:

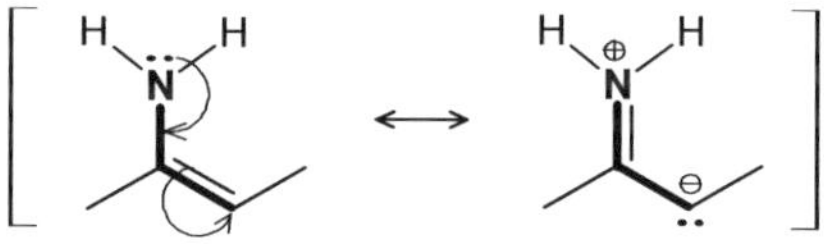

Fig. 22.6

Resonance works "both ways"

The lone pair on C is already at the very end of the area of delocalization, and the only thing it can do now, is to go back. Indeed, the double-headed arrow between resonance structures means: *resonance works both ways.* It has to.

If I showed you only the right structure, you could easily reconstruct the first, by moving the lone pair from the C atom to the adjacent bond (this is the only thing you can do at the start). It would force a π pair from the C=N bond to hop onto the N atom, and the first resonance structure would be regenerated.

We can emphasize this reversibility by drawing curved arrows on the last resonance structure. However, only few people practice this, because it is not necessary:

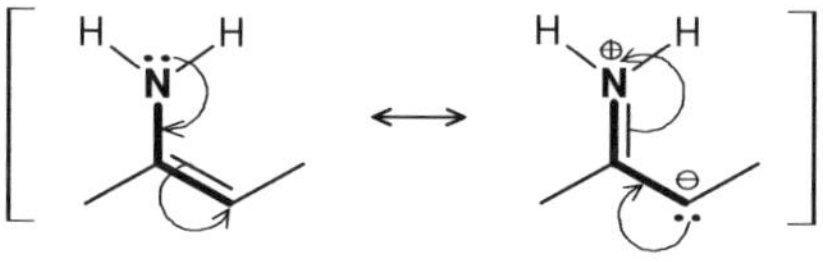

Fig. 22.7

Summary - drawing resonance structures step by step:

1 Find delocalized electrons and the area of delocalization.

2 For every delocalized pair use a curved arrow to show how you push it into the new location accessible within a delocalization area. While doing this:

- always remember that atoms in resonance structures *cannot violate the octet rule* (violation occurs whenever the number of chemically active electrons around an atom exceeds 8),
- never push any of the delocalized pairs *beyond* the area of delocalization (otherwise, you violate the octet rule instantly).

3 Having new resonance structure drawn, consider where to put formal charges:

- write ⊕ next to the atom which experienced the outflow of e^-,
- write ⊖ next to the atom, which experienced the inflow.

4 Decide whether the new resonance structure is the final one. If yes, draw a closing bracket. Otherwise, when there is some more place in the area of delocalization to push electron pairs further, draw a double headed arrow and return to step 2.

Remember, that since the delocalization concerns π and lp electrons only, σ pairs are never touched. Remember, that the overall charge of all resonance structures must be equal. Count it to check the correctness. Moreover, it is worth mentioning that there are only three scenarios for every delocalized pair, when we switch from one resonance structure to the next: (1) lone pair becomes a π pair on an adjacent bond; (2) π pair becomes another π pair on an adjacent bond. The third scenario is (3) π pair becomes a lone pair on some atom. The latter happens only when the pair is forced to (that is when there is no more place for it to become another π pair).

Problems

All of the molecules below contain delocalized electrons. For each of them, draw the set of resonance structures, to better represent molecule's nature:

22.1 OH

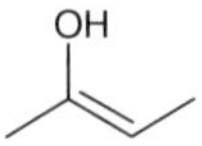

22.2 O

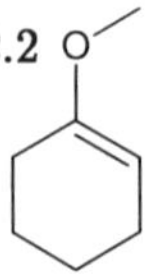

22.3 O

22.4 O NH$_2$

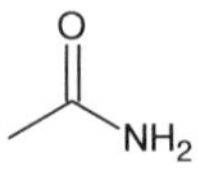

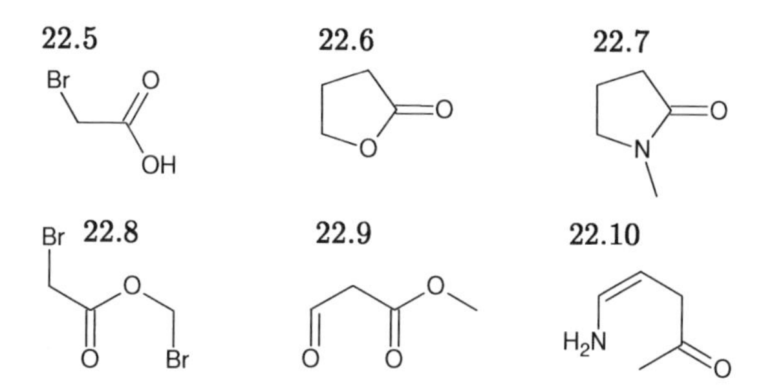

In the following molecule three pairs of electrons are delocalized together. Draw resonance structures of this amine, and note, that one of π pairs will not be forced to hop on C (the second scenario "π pair becomes another π pair on an adjacent bond" will take place):

22.11 NH_2

In the following dialkene, there are two conjugated π pairs, and no lone pairs. Try to draw its *three* resonance structures:

22.12

Molecules which contain two conjugated π pairs: one in the C=C bond, and one in the C=O or C=N bond

This is a special case. For such molecules, we will always draw *three* resonance structures. Consider the example below. It is apparent that two π pairs are delocalized together, because exactly one σ bond separates them. We treat the lone pair on the N atom as *not participating*. It is so, because its distance to the N=C π pair is 0 σ bonds, while the distance to the C=C π pair is 2 σ bonds. The lone pair is not conjugated with any of these.

We must make the first move with one of two π pairs:

H N ⟷

Fig. 22.8

Which one to choose? Remember, that in such molecules *we treat N and O atoms as sucking electrons.* The first move is to push the π

pair onto N, without touching the other – *in this case we do not move all electron pairs simultaneously!*

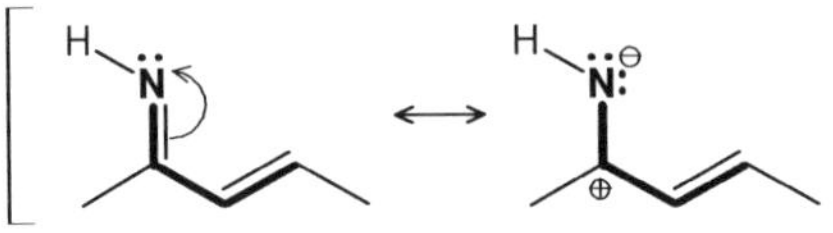

Fig. 22.9

The next move is easy to predict, when you realize that C atom with ⊕ formal charge is sucking negatively charged electrons of the π pair:

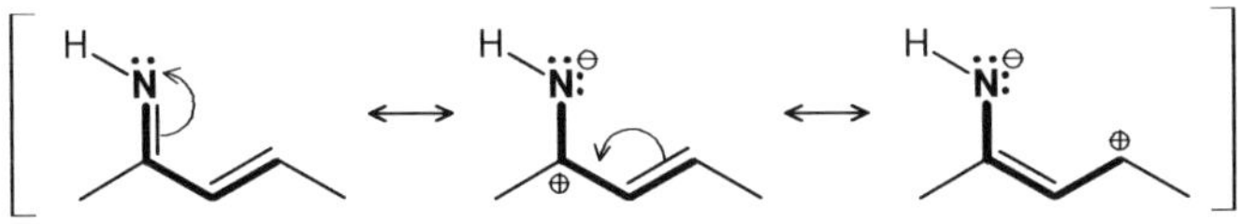

Fig. 22.10

The ⊕ formal charge on C2 disappears, because this atom experienced inflow of electrons. We decrease its formal charge by 1, so ⊕ becomes 0. On the other hand, ⊕ formal charge appears on C4, which felt outflow of electrons.

These three resonance structures are characteristic of all molecules having the C=C bond conjugated with the C=O or C=N bond. Try it yourself:

Problems

Draw the set of resonance structures for each of the following molecules, to better represent their true nature:

22.13 **22.14** 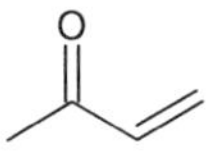 **22.15**

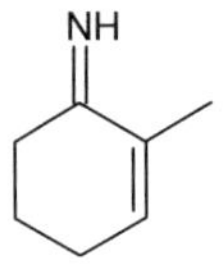

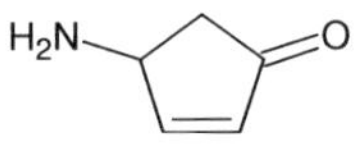

23 Formal charges of atoms in molecules

We know that *free* atoms are neutral – their charge is 0 – and that they can become charged ions. Bonded atoms, the ones within molecules, also have some charge assigned, but we call it **formal charge**.

On last pages, you have approached structure drawings with ⊕ and ⊖ symbols for the first time. ⊕ means an atom with +1 formal charge, while ⊖ means -1 formal charge. These are special situations, because normally, atoms in molecules have no formal charge at all (we can say their formal charge is 0).

Formal charges will continue to appear later in your organic chemistry course, but considering "inflow" and "outflow," as we did before, is not always comfortable. Therefore, we need a more robust method to determine formal charge of any atom in any molecule.

Assigning formal charge step by step

1 "Isolate" the atom in question from the molecule, by breaking in half all chemical bonds, which hold it.

2 Count electrons around the "isolated" atom, and assign a charge just like if it was a free atom or ion. Recall that:

- 4 e^- on C,
- 5 e^- on N,
- 6 e^- on O,
- and 7 e^- on X

ensure neutrality of the free atom.

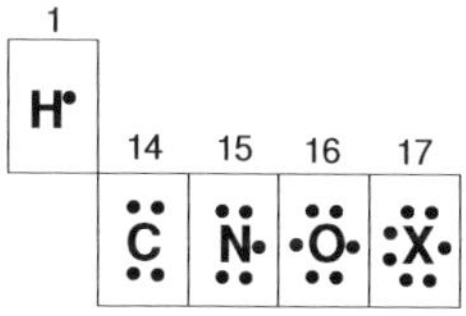

1				
H•	14	15	16	17
	C	N	O	X

Fig. 23.1

Chemically active electrons in free atoms (see: Chapter 1).

Take a look how it works for, let's say N atom in ethanamine (*not* ethan-1-amine!). The first step to assess a formal charge to the atom in a molecule is to use paper and pencil, or your imagination, in order to "isolate" the atom in question:

NH_2 ≡ N H H ⇒ N H H → N

5 e^- around
formal charge "0"

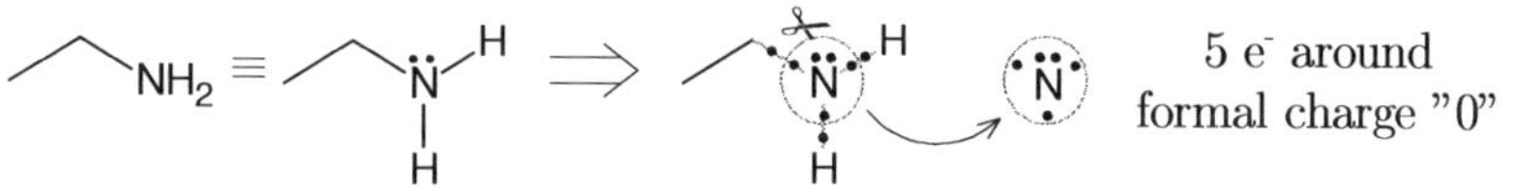

Fig. 23.2

Virtually isolated atom simulates being free. That is great, because we know how to assess charges to free atoms and ions. How many electrons isolated N atom has? Five. Simply compare the value with a known number of chemically active electrons a neutral N would have.

Since our isolated N has 5 e^- around, we can tell it is neutral. Therefore, the formal charge of the real N atom – the one within an ethanamine molecule – is 0.

Any deviation would be a sign of non-zero formal charge. For example, if some "isolated" N turns out to have 4 electrons, its formal charge in the molecule is ⊕. On the other hand, when some "isolated" O has 7 electrons, you know its formal charge must be ⊖.

Problem

Consider propynoic acid chloride and check the formal charge of all its atoms:

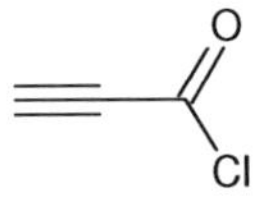

O

Cl

23.1 check Cl atom **23.2** check O atom **23.3** check C1 atom
23.4 check C2 atom **23.5** check C3 atom **23.6** check H atom

Go back to problems from the last chapter, and convince yourself, that ⊕ and ⊖ signs were correctly assigned to atoms in those structures.

24 Resonance in aromatic compounds

Resonance plays an especially important role in chemistry of aromatic compounds. Recall: aromatic compounds have molecules, which contain an aromatic ring. The benzene ring is the most significant representative of all aromatic rings:

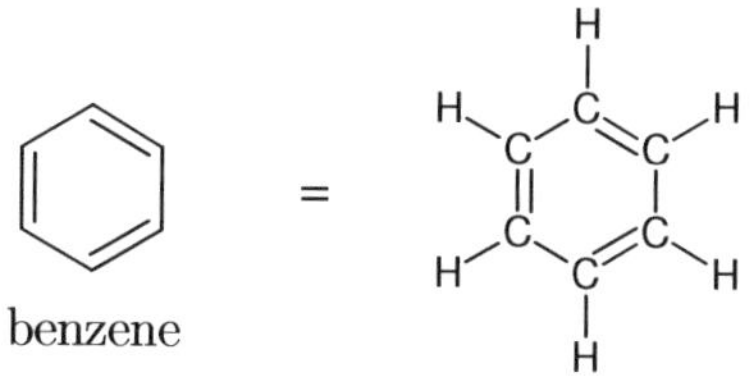

benzene

Fig. 24.1

In the chapter about functional groups (Chapter 6), I wrote, "a benzene ring does not behave similarly to alkenes." Indeed, it now appears to be *a cyclic system of delocalized electrons.*

Mind the meaning! It means we cannot say, that inside a benzene molecule, there are three C=C bonds. Such a claim would suggest that these bonds are isolated, while the situation is opposite. Three π electron pairs (six π electrons) are all delocalized. They evenly spread over a six-membered ring and belong to all C atoms equally. Benzene is a clear example of how structure drawings can mislead us.

Resonance structures of benzene

There are exactly two. They clearly show that all π electrons spread over the ring evenly, and are rather easy to draw. Just grab one of the π pairs, push it in whichever direction and you will notice that the subsequent one runs away, forcing the third to do the same:

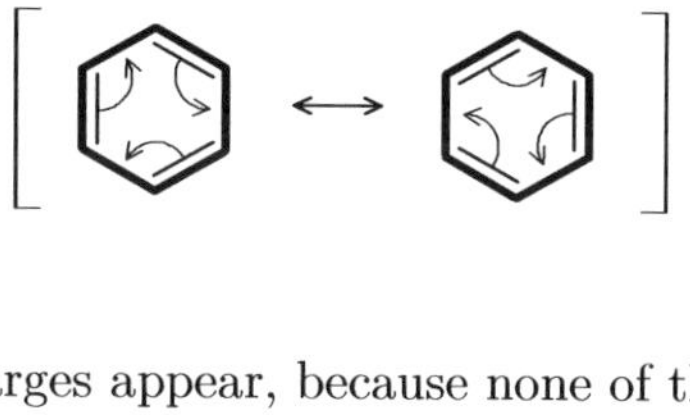

Fig. 24.2

No formal charges appear, because none of the C atoms loses or gains electrons. Moreover, we do not have to force π pairs to hop on C atoms, because the area of delocalization – which is the entire ring in fact – has no beginning nor end.

Note that it would be a tremendous mistake to push one or two pairs only, instead of pushing all of them simultaneously:

Fig. 24.3

Surprisingly, there is nothing more about the delocalization of electrons inside a benzene molecule (C_6H_6) to be expected. Maybe just one little hint: since six π electrons are all delocalized and evenly spread over the entire ring, some people write them as a circle inside the hexagon. The circle means "six delocalized electrons." You may come across such drawings in the internet or in some other textbooks:

Fig. 24.4

Benzene is just one of the thousands of aromatic compounds. There is more to expect if we broaden the scope:

Influence of substituents

In general, a benzene ring is always influenced by any substituent directly attached to it. As far as we talk about a delocalization of electrons, there are two ways to draw resonance of benzene rings with substituents:

- when the substituent is a functional group, which introduces the π pair into the delocalized system of the ring, or
- when there is a group which introduces a lone pair.

Influence of functional group introducing a lone pair to the delocalized system of the benzene ring

Look at the aromatic amine below (aniline, remember?). It is clear that the lone pair on the N atom participates in the delocalization, together with 6 π electrons of the ring:

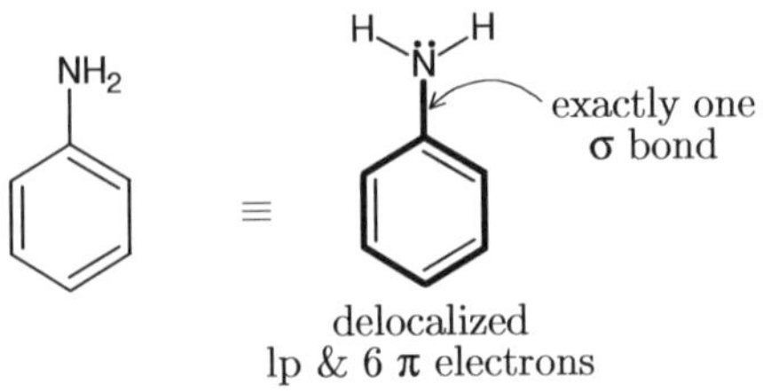

NH_2

H N H

exactly one σ bond

delocalized lp & 6 π electrons

Fig. 24.5

To draw second resonance structure of this molecule we grab the lone pair of the N atom, and push it toward the center of the area of delocalization. Therefore, we draw a curved arrow which puts it over the N–C σ bond. That incoming pair forces the first in-ring π pair to run away, and we might expect that the second and the third π pair will have to move simultaneously. After all, this is exactly what happens in a bare benzene molecule...

However, the situation is different here. Repeating the simple pattern of resonance structures, results in a disaster:

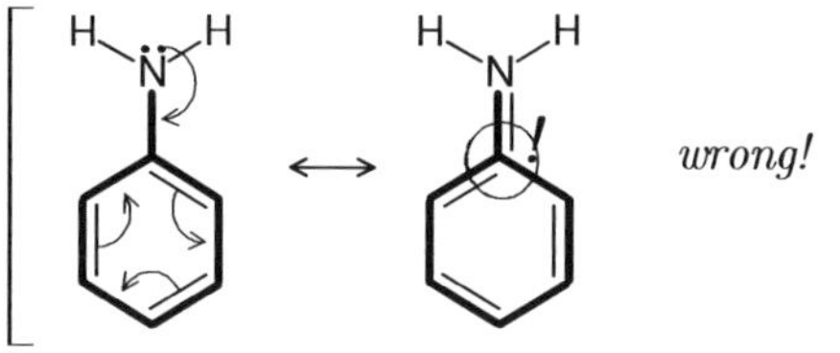

wrong!

Fig. 24.6

The snag is that we *cannot push all delocalized pairs simultaneously.* It appears that the ring becomes crowded here, and mutual movement of all pairs leads to immediate violation of the octet rule.

We had better move the first in-ring π pair onto its C atom. We force it to become the lone pair. Obviously, a ⊕ formal charge emerges on the N atom (outflow of a lone pair), while C atom gains ⊖ (inflow of a newly formed lone pair):

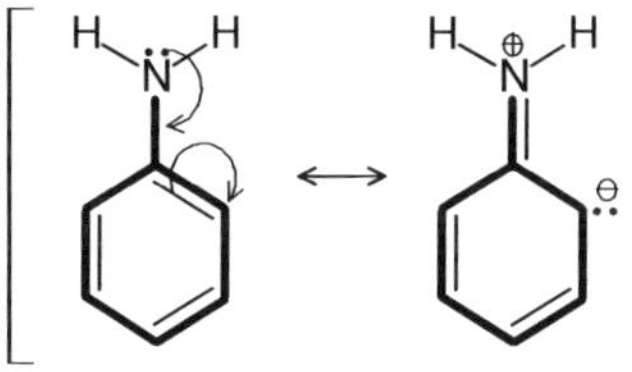

Fig. 24.7

The next step is to move the newly formed lone pair further into the delocalized area. We make it a π pair on the subsequent carbon-carbon bond, while forcing the π pair which blocks the way, to hop onto its C atom. Curved arrows on the second resonance structure show what is happening:

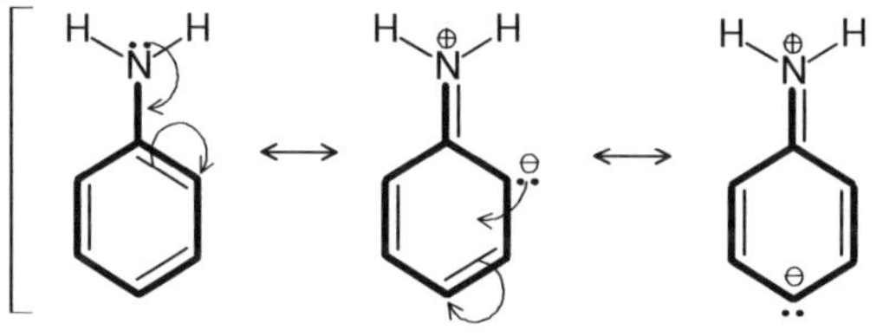

Fig. 24.8

Than we can draw the fourth resonance structure by repeating the process (arrows on the third resonance structure depict "intentions"):

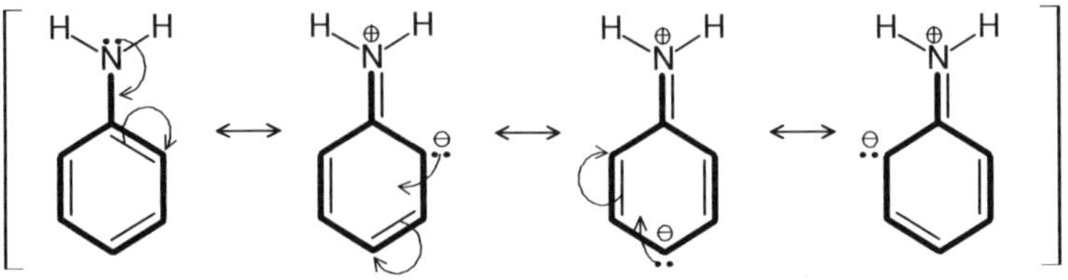

Fig. 24.9 Complete set of resonance structures of aniline.

The set of resonance structures is finished. Why? Consider the next theoretically possible move – push the lone pair onto the adjacent σ bond. It forces a π bond in C=N to hop on N and the first resonance structure appears:

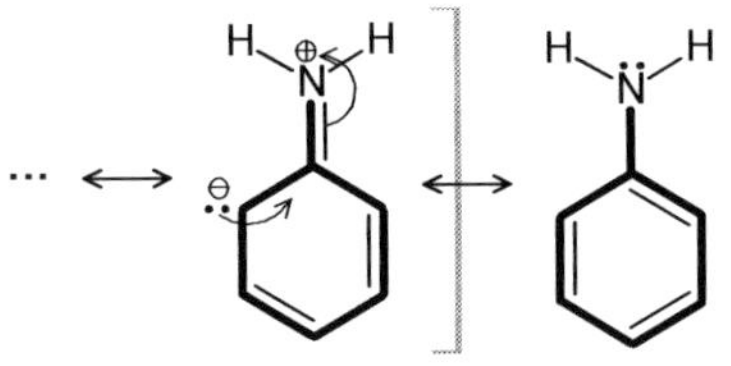

Fig. 24.10

Entire "pattern" of resonance structures will be the same for any molecule that has a functional group introducing a lone pair into the delocalized system of the benzene ring. In each of these cases, ⊖ *formal charge distributes itself evenly within the ring.*

Influence of functional group introducing a π pair to the delocalized system of the benzene ring

Now let us consider the second situation, when the group attached to the benzene ring introduces a π pair into the delocalized system. Consider the following ketone (acetophenone, remember?), in which three π pairs of the benzene ring are delocalized together with the π pair of the carbonyl group:

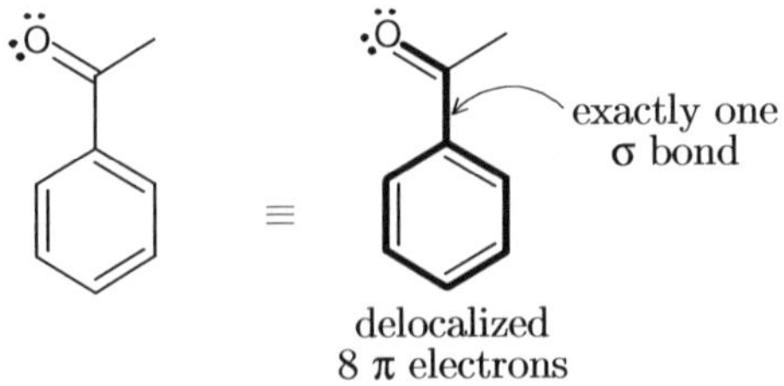

exactly one σ bond

delocalized 8 π electrons

Fig. 24.11

We saw this pattern before. Do you remember? In Chapter 22 I have devoted a separate section to "molecules with conjugated π pairs: one in the C=C bond, and one in the C=O or C=N bond."

In such cases we push a π pair in a C=O or C=N bond onto O or N atom, because O and N suck electrons from the delocalized system. Here too, it must be the first step:

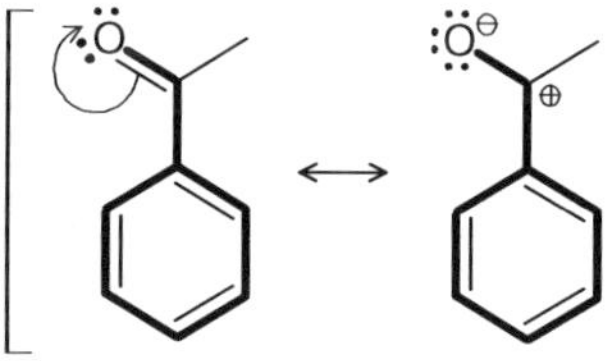

Fig. 24.12

⊕ formal charge on a carbonyl C atom attracts the first π pair:

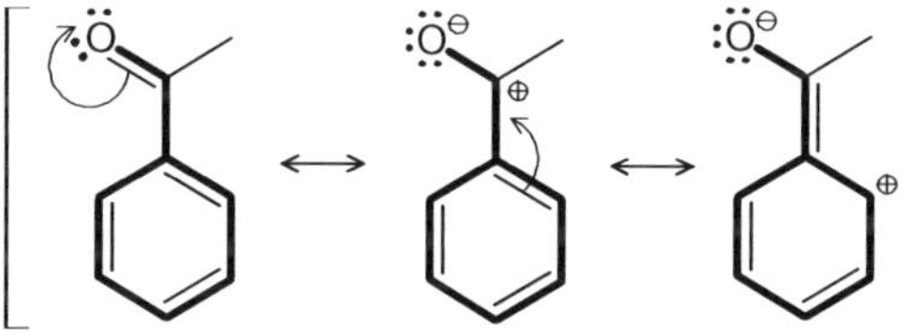

Fig. 24.13

Repeat the process twice, and you get the entire set:

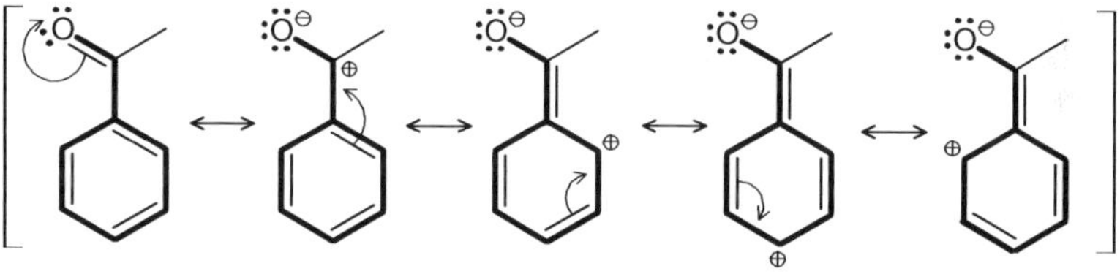

Fig. 24.14 Complete set of resonance structures of acetophenone.

The last one is indeed the final resonance structure, because the next possible move regenerates the initial drawing:

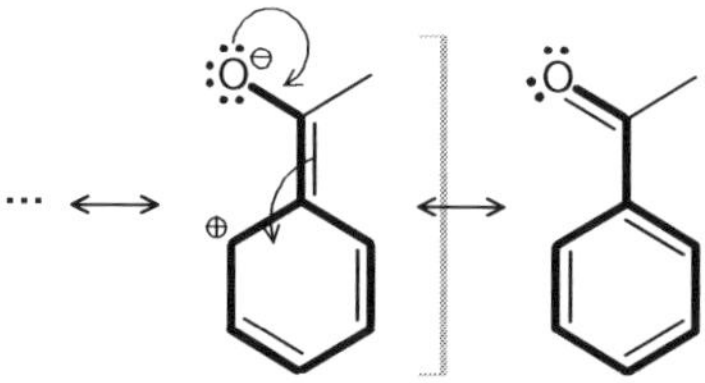

Fig. 24.15

We can sum all these up, by stating that there are two different situations in the resonance of aromatic compounds. Firstly, the benzene ring may be directly attached to a functional group, which introduces a π pair of C=O or C=N bond into the delocalized system. In such case, one of ring's π pair is withdrawn, and ⊕ formal charge appears in the ring. This charge is distributed over three C atoms.

Secondly, the benzene ring may be directly attached to a functional group, which introduces a lone pair of O or N atom into the delocalized system. In such case, the lone pair flows toward the ring, and ⊖ formal charge appears in the ring.

Both ⊕ and ⊖ formal charges in the benzene ring are always evenly distributed over three of its C atoms.

Problems

Draw the set of resonance structures for each of the following molecules, to better represent their true nature:

24.1 O Cl

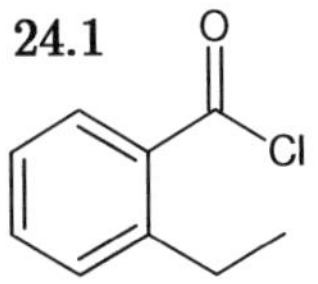

24.2 O

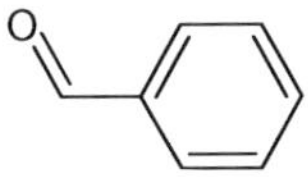

24.3 OH Br

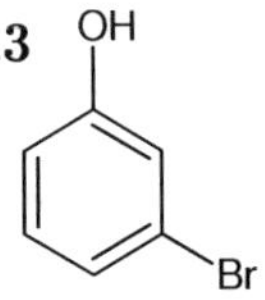

24.4 O O

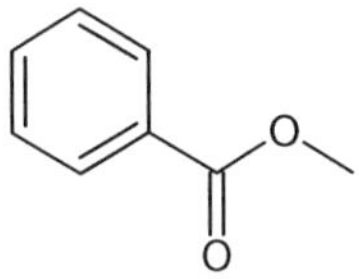

24.5 O

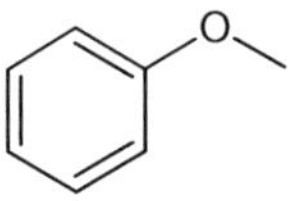

24.6 N

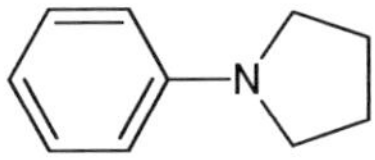

25 Understanding resonance structures

It is worth mentioning, that the term "resonance" occurs in chemistry, as well as in physics, and it has different meanings in both these sciences. Therefore, in order to avoid inconsistency, the idea appeared to eradicate the very word "resonance" from the chemical dictionary, and to replace it with **mesomerism**. Unfortunately, not all of the chemists bought the novelty, and now people call the same stuff with two competing names. We treat them as synonymous, so do not feel confused, when you come across the mesomerism in the web or textbooks.

Anyway, we can use resonance structures a.k.a. mesomeric structures – whichever you or your teacher prefers – whenever *there are* delocalized electrons within the molecule. These structures show us that electrons spread over the entire area of delocalization. The question arises: are they different "states" of the molecule, which change over time?

Resonance structures are not "states" of the molecule

Many students think so, because of the way we draw them. We use curved arrows, which suggest that electron pairs *move*. We talk about pushing electrons, and about appearance and disappearance of formal charges... The truth is that it does not happen in reality.

In fact, delocalized pairs *do not oscillate or resonate*. They spread over the area of delocalization, and it is a *completely static situation*. We have curved arrows and specific vocabulary only to help us with drawing new resonance structure, based on existing one.

In this regard, resonance structures of the molecule are not at all similar to drawings of different conformations of one molecule, which are indeed "states," which freely change over time.

Merging resonance structures into one image

To better visualize the real structure of the molecule with delocalized electrons, you should use your imagination and *impose all of its resonance structures over each other*. Such a blend is the best representation of the reality, though we do not have a unified method to draw it on paper.

Look: acetophenone has five resonance structures. The contribution of the second one is negligible, because we always treat existence of ⊕ and ⊖ on neighboring atoms as improbable. Passing over this contributor, we can visualize the real structure of an acetophenone molecule by roughly merging its four important resonance structures into one image:

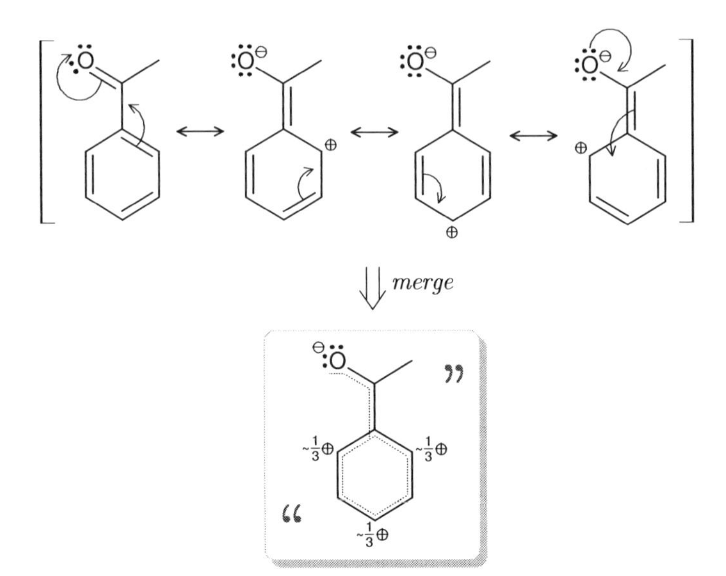

Fig. 25.1 To visualize the real nature of the molecule, we use imagination to merge its resonance structures into one image.

We see that π electrons spread over the entire σ bonds skeleton; there is constantly some negative charge on O atom, while in the benzene ring there is constantly a positive charge evenly distributed among three C atoms. All that is a static state – delocalized pairs do not move, while charged atoms remain charged all the time.

Resonance shows us that due to the delocalization, the real structure of the molecule is different, from what we see by inspecting its single drawing. For instance, would you be able to predict that three C atoms in the benzene ring of the above ketone bear some positive charge? Alas. Only by drawing all resonance structures, the feature emerges.

Sometimes in the future, resonance will turn out to be the only way to explain specific chemical properties of some molecules. The impact of resonance is especially apparent in the chemistry of aromatic compounds. Yet in this book, we will exploit it in practice, while writing acid-base reactions.

26 Electron clouds and orbitals

We are already accustomed to drawing σ and π bonding pairs as sticks, and lone pairs as dots. This redundant "system" does not provide the way to depict delocalized electrons, so we have to draw the set of resonance structures instead. Anyway, our sticks & dots are just quick to draw *symbols of localized electrons.*

In addition, these symbols are highly idealized. If chemists had agreed years ago, that the pair should be drawn as an earthworm, molecular drawings would be full of earthworms now. It is obvious that the electron is nothing like an earthworm, right? However, *symbolic sticks and dots are just as far from reality.*

Some people imagine electrons as ball-shaped, discernible particles flying around nuclei of atoms like planets orbiting a star. This is another misconception. The hypothesis was on the scene nearly 100 years ago. Times so distant that BBC radio was a novelty, and women in most countries in the world had no right to vote.

Fortunately, things change and scientific hypotheses are not an exception. Therefore, if your brain has a "solar-system-like idea of electrons" installed, it is high time to update it.

How to imagine electrons?

Think of a drip of ink. It is an easy to describe, cohesive object. Now pour it into water – the ink diffuses, becomes blurred, and loses its own individuality. Let us call what we see a cloud of ink or even better a cloud of ink density. *That cloud is made of a drip of ink, but there is no longer a drip-like object to talk about.*

Electrons in atoms and molecules are just like this. They are not ball-shaped, tangible particles, rather **clouds of electron density**, blurred around nuclei. *These clouds are made of electrons, but there are no ball-shaped particles to talk about.*

Physicists and chemists prefer to use the word **orbital** instead of "the cloud." Orbital has a deeper mathematical meaning, but in basic organic chemistry, we treat electron cloud and orbital as one. And by the way, never think of an orbital as an orbit. Is this unfortunate similarity the reason why misguided idea of a solar-system-like atom was so pervasive?

Location of electron clouds within the volume of the molecule

In most organic molecules, electrons go in pairs, and each of these pairs forms a cloud. Every σ pair is a cloud located between and around both bonded nuclei. Cloud of π pair, or two clouds of π pairs in a triple bond, surround the σ cloud.

On the other hand, we can think of delocalized electrons, as a group forming a large, shared cloud, which spreads over the entire area of delocalization. Take a look at the example below. The cloud of 10 delocalized π electrons covers the whole molecule:

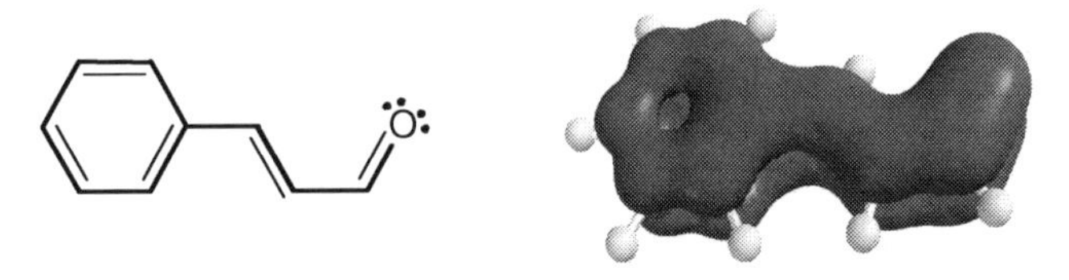

Fig. 26.1 Cloud of delocalized electrons encompasses the entire delocalization area.

This picture is a good illustration of the difference between real character of electrons and imperfect nature of the chemical drawing.

27 Induction and partial charges of atoms

In the introductory chapter, I have written that two pivotal notions in this part of the book are resonance and induction. Resonance concerns a delocalization of π/lp electrons, while induction is associated with polarization of σ bonds and appearance of partial charges on atoms inside a molecule. Let us familiarize with this new thing, which, just like the delocalization, remains invisible on structure drawings.

All atoms in a molecule bear partial charges due to induction happening through σ bonds

Free atoms, ions, and particles like electrons and protons bear an electric charge, which is an integer value. For example, a neutral H atom has 0 charge, and this zero is quite exact: 0.00000000000...

Interestingly, such *accuracy does not hold for atoms in molecules.* Why? Because atoms joined with chemical bonds are no longer free. They influence each other through the skeleton of chemical bonds. Some atoms in the molecule lose a small percent of their electron density, on the account of other atoms. I mean that electron density in all clouds out there is not evenly distributed. We refer to this phenomenon of electron density redistribution as **induction**.

Because of such delicate, though non-negligible redistribution of electron density, all atoms in the molecule are partially charged. Atom with a bit lower electron density (compared to what it would have when free), carries a small, positive partial charge, like +0.03, +0.12, or something. On the other hand, an atom with a bit higher electron density, compared to the non-bonded state, bears small, negative partial charge of, let's say -0.09, -0.14, or something like that.

Unfortunately, we cannot predict the real value of partial charges. All we can do is to determine their signs:

δ^+ symbol	denotes an atom with a positive partial charge
δ^- symbol	denotes an atom with a negative partial charge

Partial charges *vs.* formal charges

These two terms are not mutually exclusive. In fact, formal charge is only an approximation of the real value, which is the partial charge.

For instance, we know that the atom with 0 formal charge bears in fact some small, partial charge. The real value is slightly higher or lower than the approximation.

How to predict the distribution of partial charges?

We need to go back to one of high-school chemistry terms – recall that atoms differ in **electronegativity**, which outlines their ability to attract electron density from the σ bond.

Whenever two different atoms are bonded, the electron density in the bond is not evenly distributed in space. We call this phenomenon **polarization of the bond**, and refer to such a bond as **polar**. For example, in a C–H bond, C atom attracts electron density of a σ bond stronger than H, because its electronegativity is higher. Therefore, in a C–H bond, σ electron density is shifted toward C slightly. Since the cloud is negatively charged, C atom gains partial negative charge, δ^- (read: delta minus). On the opposite end of the bond, H experiences equivalent decrease of electron density, and gains partial positive charge, δ^+ (read: delta plus).

Look at the figure below, where I drew a methane molecule and added signs of partial charges. Arrowheads over bonds indicate the direction in which electron density of the bond shifts:

Fig. 27.1

Of course, the molecule, as a whole, remains neutral, because there must be perfect balance between values of all δ^+ and δ^-. Every H atom has the same value of δ^+, while δ^- on C is four times larger (it scoops electron density coming from four C–H bonds at once).

We predict signs of partial charges on two bonded atoms, by simply comparing values of atoms' electronegativity. For instance, F atoms have the largest electronegativity of all, while ability of H atoms to attract electron density is the weakest:

1	14	15	16	17
H	C	N	O	F

Fig. 27.2

Relative values of electronegativity.

Problems

Assign symbols δ^+ or δ^- to atoms in the following bonds, and draw arrow heads to show the direction of the shift of the electron density:

27.1 N–H **27.2** O–H **27.3** C–O **27.4** C=O
27.5 C–N **27.6** C=N **27.7** C≡N **27.8** C–Br

Electronegativity and polarization of bonds explain the appearance of partial charges on two bonded atoms. How does it extrapolate to the entire molecule and the phenomenon of induction?

Induction is a transfer of polarization of bonds through the entire σ skeleton

Atoms in molecules are not independent individuals. They are members of the group, joined by the net of σ and π bonds. These bonds interconnect atoms, facilitating them to influence each other, just like people influence each other through social links.

Let me show you an example. Imagine you have a triatomic molecule A–A–A. Bonds within are *nonpolar*, because every A atom has the same electronegativity. No tug of war out there, so electron density is distributed in a symmetric manner.

However, when we replace the first atom with another one, let's say B, which attracts electron density stronger, partial charges will appear: $B^{\delta-}$–$A^{\delta+}$–A. No surprise, right?

However, here comes the news: the remaining A–A bond feels polarization of B–A, and... becomes polarized too. Why? Middle A has a *partial positive charge* now, so it *attracts negatively charged* bonding pair of the A–A bond slightly. Therefore, the electron density in A–A gets shifted too! The effect is small, but not negligible. Partial positive charge on the middle A decreases, on the expense of the last A, where a small δ^+ appears: $B^{\delta-}$–$A^{\delta+}$–$A^{\delta+}$. This example means that polarization of bonds is transferred through the entire molecule, and we call it induction.

Problems

Assign partial charges δ^- or δ^+ to atoms in propan-2-one and ethanoic acid molecules. Draw arrow heads over bonds to show the direction of the shift of the electron density:

27.9

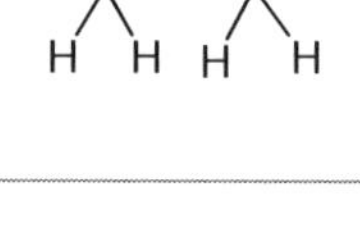

27.10

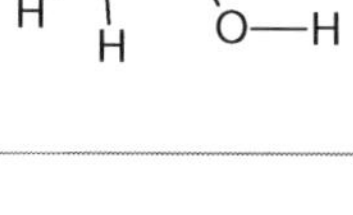

28 Electronic effect - EWG and EDG groups

Here, we can summarize this part of the book by stating that atoms and whole groups of atoms in molecules influence each other by induction and resonance.

Inductive effect

Induction is the transfer of polarization of bonds through the entire σ skeleton of the molecule. Partial charges on all atoms appear in consequence. Because in the molecule there are many mutual "influences" at once, we cannot deduce real values of partial charges.

Inductive effect of an atom, or group of atoms, or some larger fragment of a molecule is its influence on the rest of structure. Fragments, which contain electronegative atoms, *withdraw electron density inductively* from other parts of the molecule. In the same time, we can perceive fragments or groups, from which the electron density is withdrawn, as *donating electron density inductively.*

For example, the inductive effect of a F atom or a CF_3 group is that they withdraw electron density. A CH_3 group, just like all other alkyl groups, are fragments which donate electron density inductively, because electronegativity of H and C atoms is the lowest.

Resonance effect

It has a special role to play in chemistry of aromatic compounds, where we clearly see, that different functional groups attached to the benzene ring, significantly influence its electronic structure. In the ring, ⊕ and ⊖ formal charges may appear (we see them in resonance structures), depending on the **resonance effect** of the functional group.

For example, an aromatic amine will have a negative charge distributed in the ring, while an aromatic ketone has a positive charge. In both situations, it is not a result of an inductive effect, but a more important resonance effect. Simply, some groups, like carbonyl, *withdraw (from the benzene ring) electron density by resonance*, while others, like an amine group, *donate (to the benzene ring) electron density by resonance.*

Electronic effect

Electronic effect is a sum of inductive and resonance effects. In other words, electronic effect is the overall influence of the group or fragment, exerted on the rest of the molecule's structure.

EWG and EDG groups

For the sake of simplicity, in organic chemistry we like to put functional groups in the table, and describe each of them as an **electron withdrawing group** (**EWG**), or an **electron donating group** (**EDG**). The latter is sometimes referred to as electron releasing groups (ERG), but this term is less popular.

FUNCTIONAL GROUPS	INDUCTIVE EFFECT	RESONANCE EFFECT	ELECTRONIC EFFECT
–COOH –COCl –COO*C'* $–CONH_2$ –C≡N –CHO –CO	*weak* EWG	*strong* EWG	net EWG
–OH $–NH_2$	*strong* EWG	*strong* EDG	net EWG or net EDG*
–X	EWG (*strong* F > Br > Cl > I *weak*)	n/a (negligible donation of electron density)	net EWG
R– **	EDG	n/a	net EDG

*–OH and $–NH_2$ groups are EWG when only an inductive effect plays a role. When the group bonds to the benzene ring, its overall electronic effect is EDG, because resonance character dominates over the inductive effect.
**R is a symbol, which denotes *any alkyl group* (e.g., CH_3, CH_3CH_2, and thousands of others).

Problems

In the problems below, ponder on the electronic effect of groups attached to the benzene ring. Order molecules according to increasing electron density in the benzene ring:

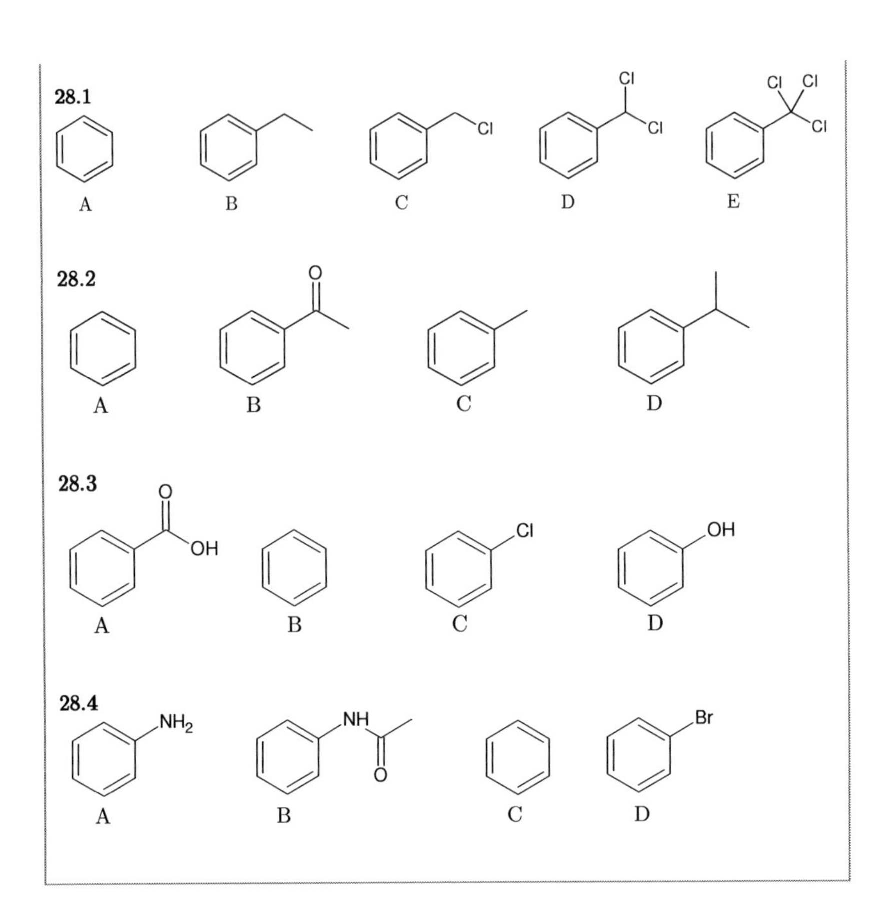
28.1
Cl
Cl
Cl
Cl
Cl
Cl
A
B
C
D
E
28.2
O
A
B
C
D
28.3
O
OH
Cl
OH
A
B
C
D
28.4
NH2
NH
O
Br
A
B
C
D

Part VI ACID-BASE REACTIONS

29 Introduction to organic reactions

Organic reaction is a *transformation of the structure of an organic molecule.* We call the starting molecule a **substrate**, and a newly formed one is the **product** of the reaction. The change of the structure always relies on the reorganization of electron pairs / chemical bonds – new ones are created, while some others are being broken.

Four basic types of organic reactions

In most reactions, the structure of the molecule changes together with its atomic composition. Whenever a product contains some new atoms incorporated, we call such reaction an **addition**. When some atoms are torn out, the product is lighter than the substrate, and we call in an **elimination**. In **substitution** reactions some fragment of the substrate molecule is being replaced with another fragment.

Nevertheless, there are reactions, in which no change of atomic composition is observed. It is the case when the transformation takes place "inside" the molecule – nothing is being added, eliminated nor substituted – only connections between atoms are changed. We refer to such reactions as **rearrangements**.

Irreversible and reversible reactions

We write organic reactions down in form of **chemical equations**. They consist of the drawing of the substrate molecule and the drawing of the product molecule, separated by one of two reaction arrows:

substrate $\longrightarrow$ product (irreversible reaction)

substrate $\rightleftharpoons$ product (reversible reaction)

Normal arrows indicate **irreversible reactions**, which rely on one-way transformation of the substrate into the product. We do not know how much time is needed, but eventually all of the substrate molecules will transform into the product. Irreversible reactions lead to the product completely, and the reversed process – from product to substrate – never occurs in the same time.

Double arrows indicate **reversible reactions**, in which products unavoidably transform back into substrates. Since there are two competing "half reactions" – from left to right, and from right to left – we will never obtain 100% yield of the product molecule. After some time the process reaches a so-called **state of equilibrium**. Both reactions are still live, but frequency at which substrate molecules give product molecules, becomes equal to the frequency of reversed process, so that no net change is observed.

The mechanism of organic reaction

Chemical equation is a simplified description of a reaction, which tells us what molecule we have at the start, and with what kind of molecule we finish.

However, we often use elaborate equations, called **mechanisms of reactions**. The mechanism tells us what is really happening down there – it shows, step by step, all changes occurring to bonding pairs and lone pairs during the transformation. Many reactions have intricate, multi-step mechanisms. Others, like mechanisms of acid-base reactions, are very simple.

In the mechanism, we use **curved arrows**, like in the resonance. Here, the meaning is different, however. In resonance, curved arrows helped us to draw subsequent resonance structures, and were not indicators of any physical phenomena. In organic reaction mechanisms, curved arrows point out the real changes electron pairs experience. They show from where to where electrons go, when new bond forms, or old one breaks.

From organic molecules to organic reactions

A basic organic chemistry course can be divided into two sections: molecules and reactions. This book is devoted to the first – we discuss *organic molecules*, and learn features determining their behavior. Understanding molecules is a prerequisite for a successful development of your skills. It will make your later studies in *organic reactions* efficient, so that you have a chance to perceive it as a rewarding challenge, rather than a painful grind.

Although the book is about organic molecules, in this very part we will familiarize with the first type of organic reactions. They will show you how resonance and induction effects work in practice.

From organic reactions to organic synthesis

The third degree of initiation is **organic synthesis**. Usually academic teachers expound on this branch of organic chemistry during advanced courses. Nevertheless, you will have to familiarize with the basics.

Organic synthesis is how most organic chemists make money to live. When you know and understand really a large number of organic reactions, you can do the job. It involves a lot of thinking and creativity, and the tasks are as follows:

- *To work out and perform a new synthetic pathway, which leads to a known and important molecule.* This is a case whenever some compound is selling for decent price, and we believe that there must be some better and cheaper way to produce it. The pursuit is driven by someone's willingness to, for example, make the process more "eco-friendly." However, finding a cheaper synthetic pathway is financially rewarding.

- *To work out and perform the synthesis of a molecule that no one has ever made.* This is a case when we design a new molecule on paper, and there is a conjecture that specific properties and practical applications are to be expected. The pursuit is driven by someone's willingness to develop human scientific knowledge, and earn some money in the future.

- *To work out completely new organic reaction.* It means finding unknown synthetic connections on the map of organic reactions. New types of organic transformations may be applied to synthesis of new chemical molecules (*task 1*), as well as to finding new synthetic pathways (*task 2*). The pursuit is driven by curiosity, but sometimes there is some money on the horizon.

In this part of the book I'm going to show you how acid-base reactions work – this subject is a classic introduction to organic reactions. No one makes money from it, but even Brad Pitt started his career as a chicken mascot in a fast food restaurant.

30 Organic bases and protonation reaction

Molecules often trade with H^+ cation. This is a regular player in organic chemistry and people call it a **proton**. Why? Look: neutral H atom is made of 1 p^+ and 1 e^-, so when we tear the electron out of it to get H^+, what is left is a mere p^+. Therefore, symbols "H^+" and "p^+" mean the same:

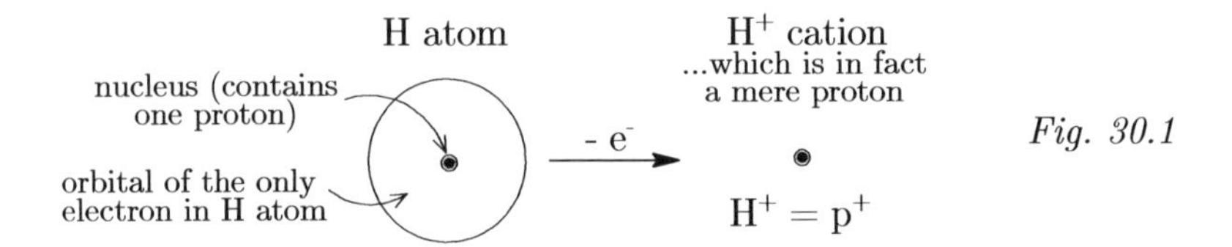

Fig. 30.1

Recall from high school chemistry, that **acid** is every molecule, which can *release a proton*, while **base** is every molecule, which can *bond a proton*. We call *elimination* of H^+ from an acid molecule a **deprotonation** reaction, and refer to *addition* of H^+ to the base as **protonation**.

When acid and base are mixed in one reaction flask, deprotonation of the acid occurs, followed by protonation of the base. Chemists refer to this blend of two reactions as **acid-base reaction**. Most acid-base reactions, as well as protonation and deprotonation, are reversible reactions, so we will use double arrows in their equations.

Two features make these transformations important. Firstly: educational potential. Acid-base reactions are great to learn how to use resonance and induction in practical problems. Secondly: abundance. They are very "popular," and often work behind the scenes. In the future, you will discern (de)protonations sometimes, as humble, elementary steps in mechanisms of more elaborate organic reactions.

In this chapter, we will discuss details of protonation of bases, while the next one is generously devoted to deprotonation of acids. Chapter 33 merges these two simultaneous processes into acid-base reaction.

How is it possible for protonation to occur?

Protonation is adding the proton to the molecule. By adding, we mean *the formation of a new chemical bond* between H^+ coming from the outside, and one of atoms within the molecule. But the chemical bond is made of electron pair, right? So how is this process possible, if H^+ has literally no electrons to share?

Simply, both electrons we need must come from the molecule. Where do we get these? We know that there are chemically active electrons in σ bonds, π bonds and lone pairs. The answer turns out to be simple: lone pairs are the choice; since only these are not yet engaged in bond formation (they are not so useless, as you see!).

Lone pairs appear on N, O and X atoms, and they significantly differ in the ability to bond protons. Why? Electronegativity has a role to play – *the more electronegative the atom is the stronger it holds its lone pairs.*

GENERAL ABILITY OF LONE PAIRS TO BOND PROTONS	
lone pairs on X	*never bond protons* Halogens are either very electronegative (like fluorine), holding their lone pairs so tightly that they become immobilized; or huge (like iodine), making their lone pairs inaccessible from outside.
lone pairs on O	*bond protons unwillingly* Oxygen has high electronegativity, so it holds its lone pairs tightly, but not as tight as X. It can become protonated under special conditions – when there are really a lot of protons around (we say, that the environment is highly acidic).
lone pairs on N	*are best in bonding protons* Nitrogen has the lowest electronegativity among N, O, X, and bonds protons quite easily, even when there are not too many protons around. That is why typical organic bases are molecules, which contain nitrogen functional groups.

How to draw protonation reaction?

Protonation occurs when a lone pair catches a proton, and becomes a σ bond. According to the table above, lone pairs on N are best in this business. Therefore, let's use protonation of ethanamine as an example:

$$\text{CH}_3\text{CH}_2\text{NH}_2 + \text{H}^{\oplus} \rightleftharpoons \text{CH}_3\text{CH}_2\overset{\oplus}{\text{N}}\text{H}_3$$

Fig. 30.2

This is normal chemical equation. However, we can use a curved arrow to show the **mechanism** of this one-step reaction. It is very simple, and for all protonation reactions, it will always be like that:

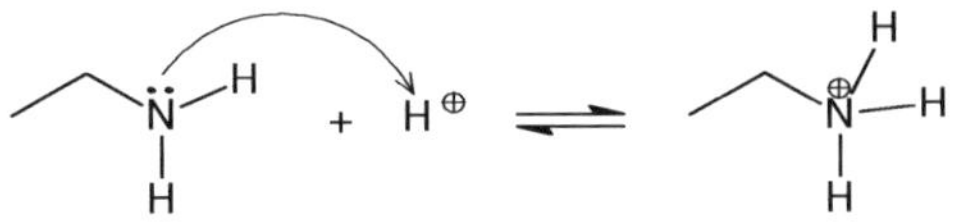

Fig. 30.3

According to the mechanism, and according to what tail and head of the curved arrow show, the lone pair attacks the proton and bonds it.

The product contains *a new functional group:* NH_3^+, which has a ⊕ formal charge located on N. In the course of reaction, formally neutral N shares its lone pair. It makes the N atom positively charged, because it feels an outflow of electron density. Meanwhile, the proton becomes a normally looking, neutral H.

From a distance, it looks like ⊕ hopped from H to N. You may use our robust method to assign formal charges to check, that there is indeed a ⊕ on N. This group, NH_3^+, looks a bit strange. However, it is totally fine. Note that N atom does not violate the octet rule. It looks unfamiliar, because we see it for the first time. Not for the last, however, because ionized functional groups are hallmarks of all acid-base reactions.

Molecular ions

Products of protonation have an increased overall charge, because ⊕ "hops" from free H^+ to the atom in the molecule. Thus, when the substrate is a normal, neutral molecule, the product is a cationic structure. Something new, right?

Although all molecules, you have seen so far in the book were neutral, this is not the rule. Indeed, so-called **molecular ions** are quite common, especially in living organisms. Moreover, all of them are products of protonation (this chapter) or deprotonation reactions (next chapter).

Cationic product of protonation is always an acid

Now, let's put an emphasis on something else. You should realize that the cationic product of protonation reaction is an *acid*. As I have written at the beginning of the chapter, "acid is every molecule, which can release a proton." Now, look at the above reaction, but read it "from right to left." Or take a look at a rewritten version:

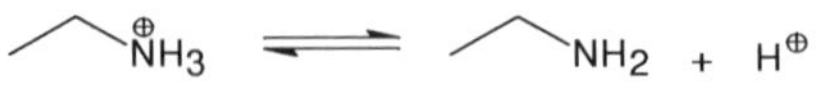

Fig. 30.4

It is apparent that in the reversed reaction the cationic product releases the proton. It must be an acid, by definition.

By the way, some people refer to such acid-base pair as conjugated. We can say that the cationic product is a **conjugate acid** of the base used as a substrate.

Problems

All molecules below are products of protonation reactions. Please, check the formal charge of every N and O atom (use our strict method from Chapter 23, which relies on "isolating" them from their molecules). Add signs of formal charge (⊕) wherever necessary:

30.1 30.2 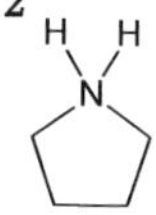 30.3

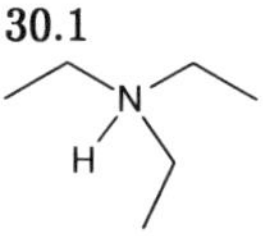

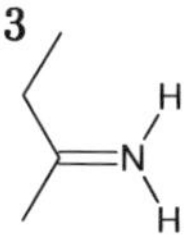

30.4 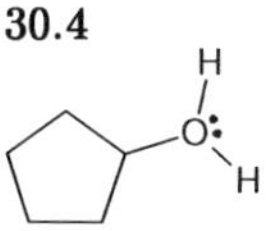 30.5 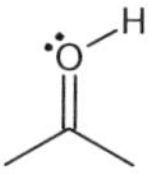 30.6

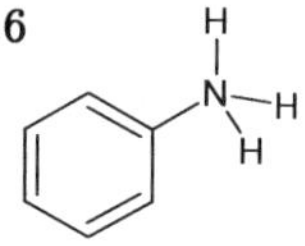

Write mechanisms of reactions, which lead to these molecular cations (you must deduce the structure of the substrate molecule). In each of equations, mark every structure as an "acid" or "base":

30.7 ? ⇌

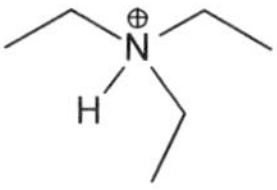

30.8 ? ⇌

30.9 ? ⇌

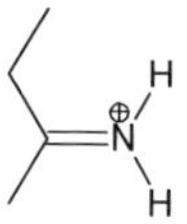

30.10 ? ⇌

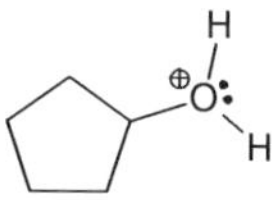

30.11 ? ⇌

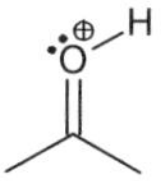

30.12 ? ⇌

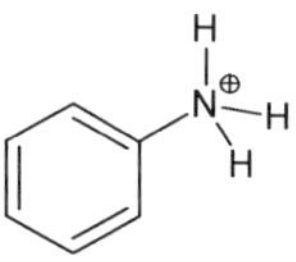

Write *mechanisms* of protonation of following molecules. Assuming the ratio of one proton per one molecule, you must predict which of the lone pairs bonds the proton. Sign all structures as "acids" and "bases":

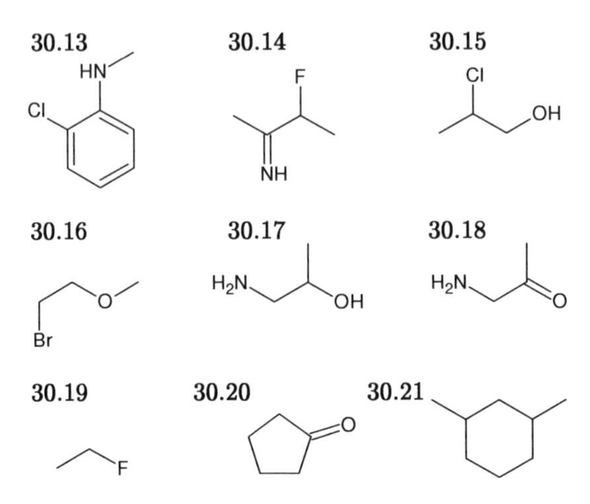

30.22 Taking into account the type of atom (N/O/X), group molecules **30.13-30.21** according to descriptions:

A – relatively good base, bonds the proton easily,
B – very weak base, bonds H^+ unwillingly,
C – a non-basic molecule, which does not bond the proton.

31 Organic acids and deprotonation reaction

How is it possible for deprotonation to occur?

Deprotonation reaction is an example of elimination – a proton is uncoupled from the molecule. To uncouple it we must break the bond, which joins the proton with some other atom in the molecule, like N or O.

The proton flies away with no electrons at all, so what happens to the σ pair? Well, the pair cannot vanish. *Both electrons must land on the atom in the molecule as its new lone pair*, when H^+ frees itself.

Atoms that bond H in organic molecules are C, N and O; therefore, we are confined to three types of bonds: C–H, N–H or O–H. Once again, electronegativity of atoms will help us to understand differences in their chemical properties. In general, *the more polarized the bond, the easiest it is to break it with H^+ release*:

General ability of bonds to be broken with H^+ release	
O-H bonds	*are easiest to break with H^+ release* O–H is the most polarized bond of all three. It means that the bonding pair is already quite shifted toward O. Therefore, breaking the bond, that is moving the pair onto O, and releasing H^+ with no electrons at all, must be the easiest in this case.
N-H bonds	*are resistant to be broken with H^+ release* The N–H bond is less polarized than O–H. The density is more evenly distributed between N and H atoms. Therefore, it is much harder to push it, as a whole, toward N atom, in order to release free H^+.
C-H bonds	*are extremely resistant to be broken with H^+ release* The C–H bond is a very weakly polarized bond, so it is very resistant to become broken. However, we cannot treat C–H bonds as totally unbreakable, because the process may occur under some special conditions (you will learn about it only in later parts of the organic chemistry course).

How to draw deprotonation reaction?

According to the table above O–H bonds are easiest to break with H^+ release, so let's use deprotonation reaction of methanol molecule (CH_3OH), as an example:

Fig. 31.1

The curved arrow in the mechanism above, indicates that the bond between O and H atoms breaks – the pair lands on O, and H uncouples as a proton. It does not take away any electrons.

The product is a molecular anion, which contains *a new functional group:* O^-. This group is characterized by the presence of ⊖ formal charge on O. The atom experienced inflow of electron density, while H atom felt outflow, thus gained ⊕.

Anionic product of deprotonation is always a base

As I wrote at the beginning of the previous chapter, "base is every molecule, which can bond a proton." Now, look at the above reaction, but read it "from right to left." It becomes apparent, that in the reversed reaction the anionic structure bonds the proton. It must be a base, by definition. By the way, some people call it a **conjugate base** of the acid used as a substrate.

Problems

All molecules below are products of deprotonation reactions. Please, check the formal charge of every C, N and O atom. Use the strict method from Chapter 23. Add ⊖ sign wherever necessary, and remember that one ⊕ will also be needed:

31.1

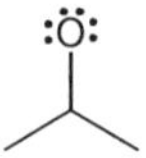

31.2

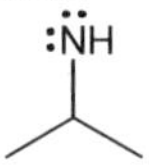

31.3

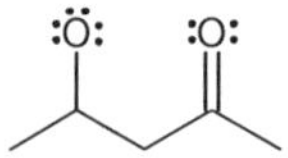

31.4

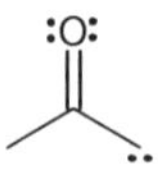

31.5

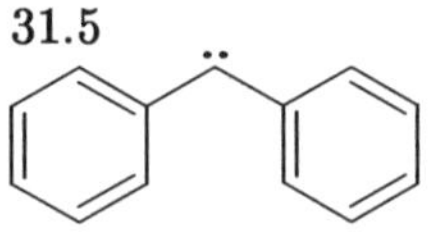

31.6

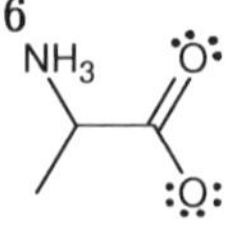

Write mechanisms of reactions that lead to these structures (you must start from deducing the structure of the substrate molecule). Also, in each of equations sign structures as "acid" or "base":

31.7 ? ⇌ :Ö:⊖

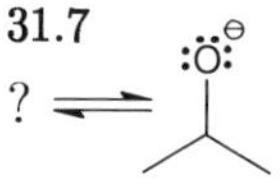

31.8 ? ⇌ ⊖:NH

31.9 ? ⇌ :Ö:⊖ :O:

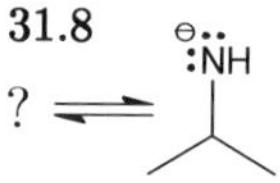

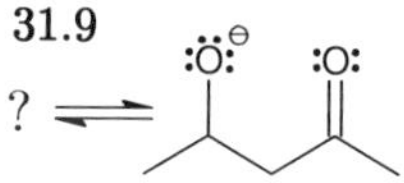

31.10 ? ⇌ :O:

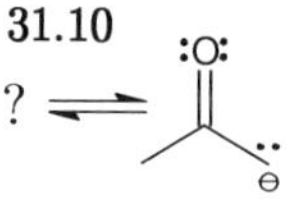

31.11 ? ⇌ ⊖

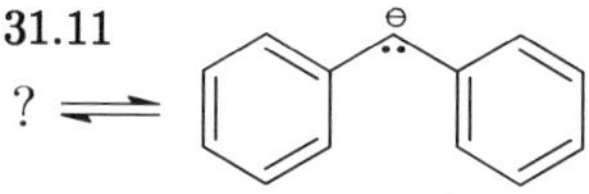

31.12 Take a look at the zwitterion **31.6**. The process in which it forms is a two-step transformation. At first, one group deprotonates, and then the released proton reacts with another functional group. Try to write mechanisms of these subsequent reactions.

31.13 In the substrate you have drawn in problem **31.12**, the NH_2 group may come close to COOH. It happens in a particular conformation of this molecule. Then, there is a chance for the proton to hop from COOH to NH_2 *in one step*. Try to write the mechanism of such a process.

Write mechanisms of deprotonation of following molecules. You must remove one proton only, so your task is to decide which one is that – which one is the most acidic. Also, in each of equations sign structures as "acid" or "base":

31.14 HN Cl

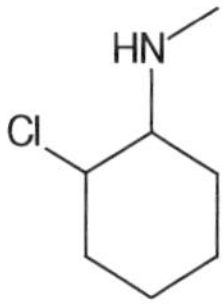

31.15 F OH

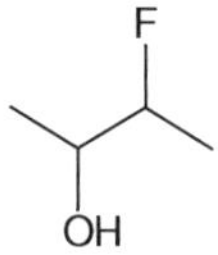

31.16 NH_2 OH

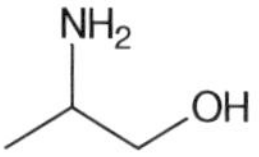

31.17 OH NH_2 O

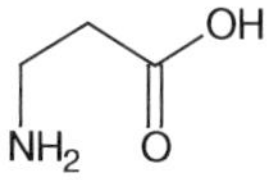

The rule of charge conservation

It says that *in every transformation the overall electric charge remains unchanged.* It was useful in resonance, and is true for organic reactions too.

The overall electric charge of all species on the left side of the chemical equation must be equal to the overall charge of species on the right side.

You can always apply the rule to check the correctness of your chemical equations. Indeed, sometimes it happens, that we lose a charge, or add one too many. Count the overall charges on both sides – whenever they are not equal, your equation cannot be correct:

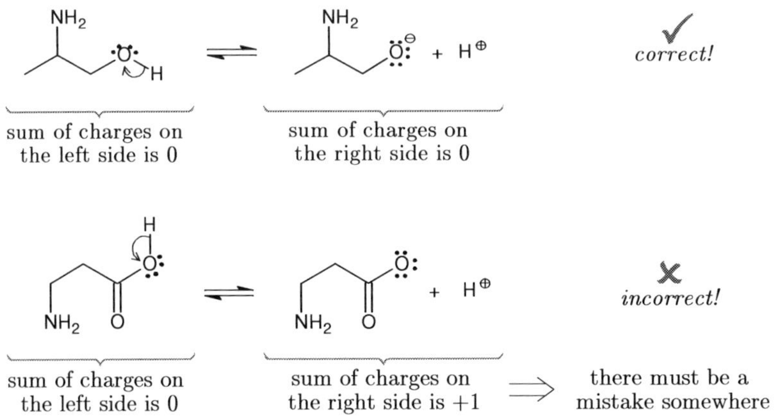

Fig. 31.2

32 Comparing the strength of acids and bases

Writing equations of protonation and deprotonation reactions is quite simple. However, in acid-base chemistry the snag is somewhere else. The true challenge is to *compare strength of various bases*, and to *compare strength of various acids*. And we do it using our knowledge about resonance and inductive effects.

Neutral and ionic acids and bases

After two previous chapters, we can distinguish four kinds of players in acid-base chemistry. Two of them we have met in Chapter 30...

(1)	neutral molecules acting as bases	$\underset{- H^-}{\overset{+ H^+}{\rightleftharpoons}}$	cationic structures acting as acids	(2)

...and the other pair in Chapter 31:

(3)	neutral molecules acting as acids	$\underset{+ H^+}{\overset{- H^-}{\rightleftharpoons}}$	anionic structures acting as bases	(4)

Of course, there are many organic molecules that are unamused by acid-base chemistry. They exhibit neither basic nor acidic properties.

Anyway, now I am going to show you how we compare the strength of various bases, and the strength of various acids. Interestingly there is a single method to do it – we look at a base (1) or anionic base (4) and deduce the willingness of its lone pair to bond the proton. Believe me or not, but this single method will solve all problems, including comparison of the strength of acids.

Comparing the strength of bases

Bases are molecules that use their lone pair to bond a proton. Therefore, we focus on the lone pair's willingness to do it, because it is a direct gauge of basicity. The more reactive the base is, the higher its so-called **strength**. Our task is to ponder on how *resonance and inductive effects influence the strength, by modifying lone pair's reactivity toward proton.*

Resonance effect significantly influences basicity – if the lone pair is delocalized, it is much less reactive toward the proton.

Delocalized pair does not sit in place, patiently waiting for a proton to catch. Also, you can imagine that it is hard for a proton to find it, because delocalized pair is blurred over the area of delocalization.

Moreover, some resonance structures of the molecule, which contains an atom with a delocalized pair, will have a ⊕ formal charge on that atom. It is not encouraging for H^+ to come closer, because two positive charges always repel each other.

The influence of the resonance effect on the strength of bases, is clear. Take these two amines as an example:

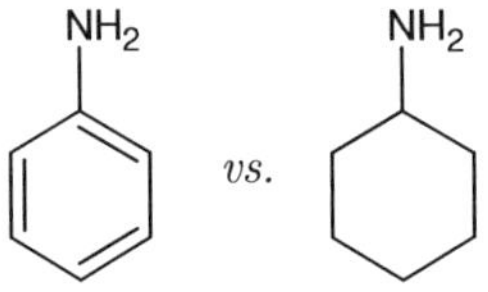

NH_2 *vs.* NH_2

Fig. 32.1

Left amine (aniline) is weaker, because its lone pair is delocalized (go back to its resonance structures: Fig. 24.9). Therefore, the lone pair is less reactive toward the proton, when compared with a localized pair of the right cyclohexanamine (*not* cyclohexan-1-amine).

Now, try to compare the strength of two anionic bases:

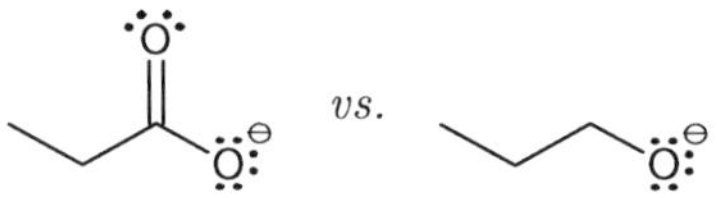

vs.

Fig. 32.2

Existence of a negative formal charge is a good thing, because it additionally attracts the proton. It is especially significant in the right anion (product of deprotonation of butan-1-ol), which is a strong base.

Left one (product of deprotonation of propanoic acid) is a weaker base, because the lone pair is delocalized. In addition, the negative charge spreads over two oxygen atoms, so that its attractive attitude diminishes in eyes of a positively charged proton:

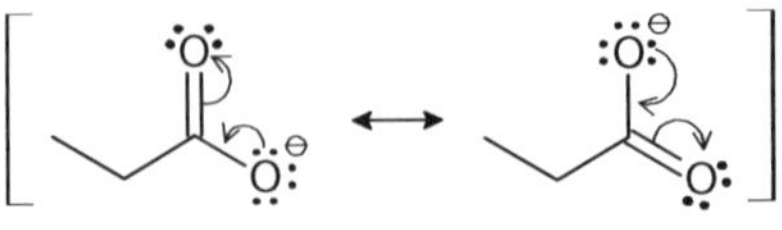

Fig. 32.3

Inductive effect also significantly influences basicity – a lone pair on an atom surrounded by groups that withdraw electron density inductively, is less reactive toward the proton.

Why? Because the atom with the lone pair is more electron-deficient, and its partial charge is decreased. In consequence, the atom holds its lone pair more tightly.

In the absence of EWG groups, the electron density of the atom is higher, thus it is more ready to share its lone pair. Take a look at examples:

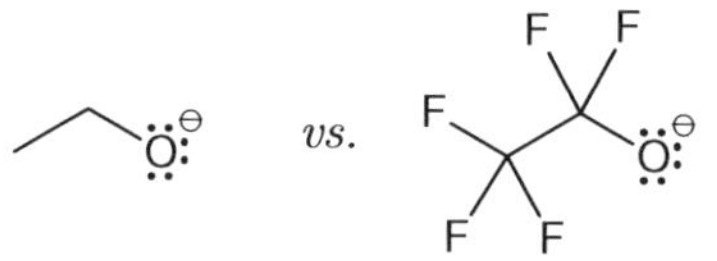

vs.

Fig. 32.4

The left anion is a stronger base. We always classify alkyl groups, like an ethyl group (CH_3CH_2), as electron donating groups (Chapter 28). In other words, electronegative atoms, like N and O, suck their density with ease, because both C and H atoms have low electronegativity. Therefore, O^- in $CH_3CH_2O^-$ is electron-rich, and more willing to share one of its lone pairs with the proton.

In the right molecule situation is quite the opposite. F atoms are highly electronegative, so the entire CF_3CF_2 group acts as electron density "withdrawer." It sucks density from O^-, diminishing its willingness to give up the lone pair. It all works well for neutral bases too. Look:

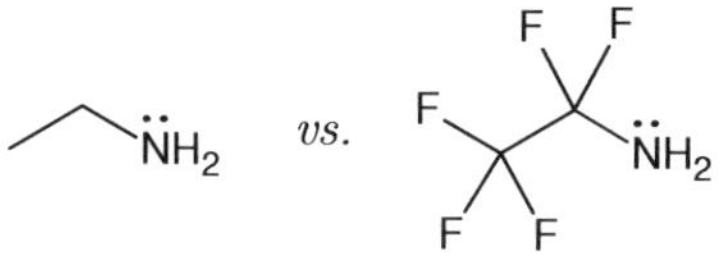

vs.

Fig. 32.5

Left one is much stronger. Or consider yet another example:

vs.

Fig. 32.6

Due to the inductive effect, the right amine is stronger, than the left one. Alkyl groups donate electron density (in other words: N sucks their electron density). Since in the right amine there is not only one, but two of them, the right amine must be a stronger base.

Problems

Determine which of two molecules is a stronger base – which is more reactive toward the proton? Explain each of your choices (there are explanations in answers).

32.1 *vs.*

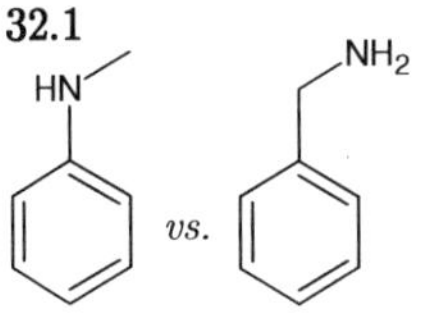

32.2 *vs.*

32.3 *vs.*

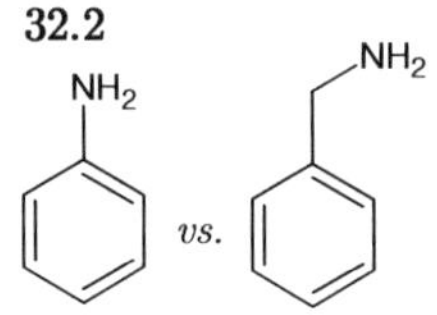

32.4 *vs.*

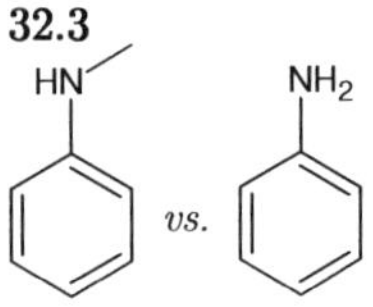

32.5 *vs.*

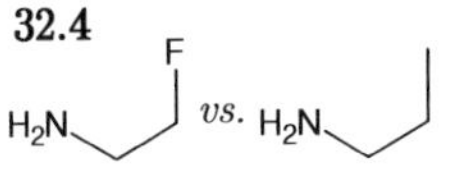

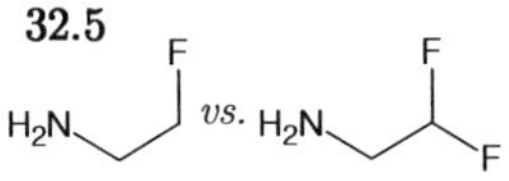

32.6 *vs.*

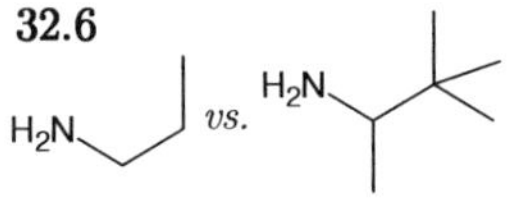

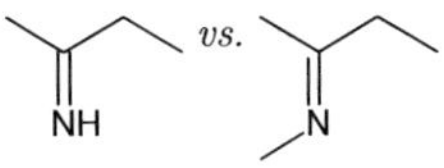

32.7 *vs.*

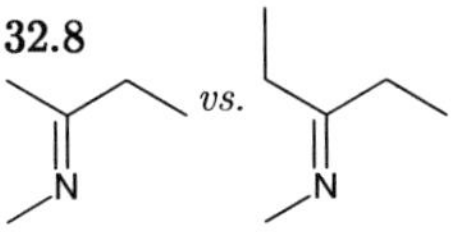

32.8 *vs.*

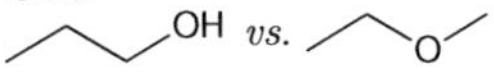

32.9 *vs.*

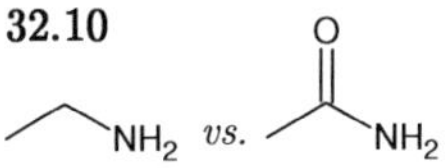

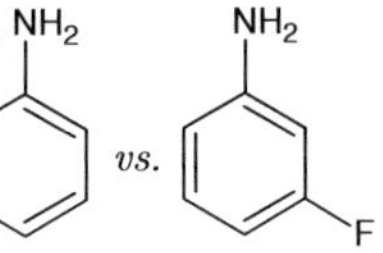

32.10 *vs.*

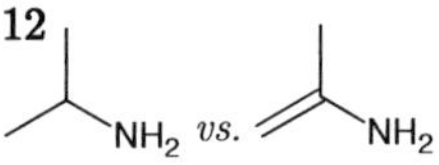

32.11 *vs.*

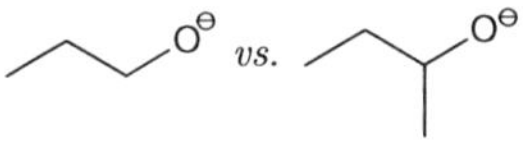

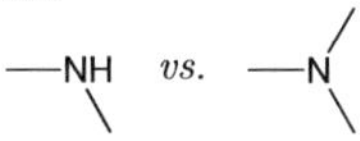

32.12 *vs.*

32.13 *vs.*

32.14 *vs.*

32.15 *vs.*

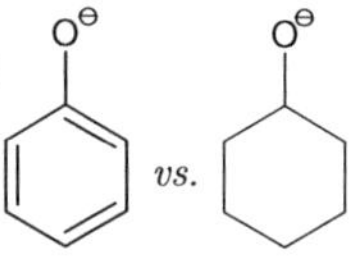

32.16 Draw resonance structures of the aromatic anion **32.15**, to prove that it is a weak base.

Comparing the strength of acids

Comparing the strength of acids *relies on drawing structures of corresponding conjugate bases*, and comparing their strength, just as we did above. And then we use the rule of thumb:

when the conjugate base is relatively strong, the starting acid is weak
when the conjugate base is relatively weak, the starting acid is strong

For example, we can compare acidity of two alcohols: CH_3CH_2OH and CF_3CF_2OH, by considering basicity of their conjugated bases: $CH_3CH_2O^-$ and $CF_3CF_2O^-$. Because of the inductive effect, $CF_3CF_2O^-$ is a weaker base than $CH_3CH_2O^-$. Therefore, corresponding acid CF_3CF_2OH is stronger than CH_3CH_2OH.

Problems

For each of pairs below, draw structures of corresponding conjugate bases. Then tell which of two acids is stronger, and explain your choices.

32.17 OH *vs.* OH

32.18 OH *vs.* OH F

32.19 O NH$_2$ *vs.* NH

32.20 F H$_2$N *vs.* H$_2$N

32.21 F HO *vs.* F HO

32.22 HO *vs.* HO

32.23 HO O *vs.* HO O

32.24 HO O *vs.* Cl HO O

32.25 HO O *vs.* HO O

32.26 ⊕NH$_3$ *vs.* ⊕NH$_3$

32.27 ⊕NH$_3$ *vs.* ⊕NH$_2$

32.28 ⊕NH$_3$ *vs.* ⊕NH$_3$

33 Acid-base reactions and equilibrium

Look at two redrawn reactions – protonation of ethanamine and deprotonation of methanol, which I have used as examples before:

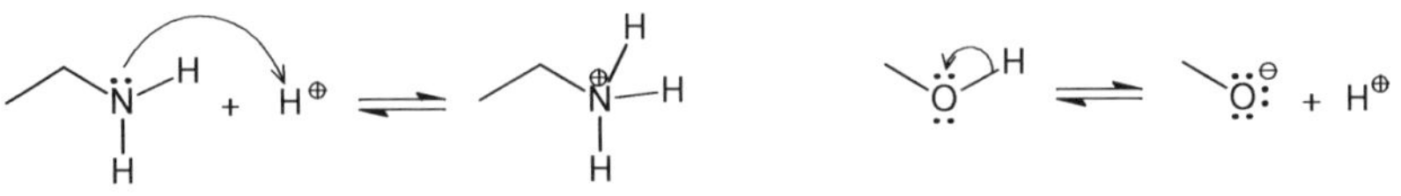

Fig. 33.1

Where does the proton reacting with an amine come from? Where does the proton eliminated from an alcohol go?

I have written in Chapter 30 that "deprotonation of an acid and protonation of a base occur simultaneously, and the blend of these two reactions is referred to as **acid-base reaction**." So this is it – both reactions can occur simultaneously in one reaction flask, when ethanamine and methanol are mixed. The corresponding acid-base reaction looks like this:

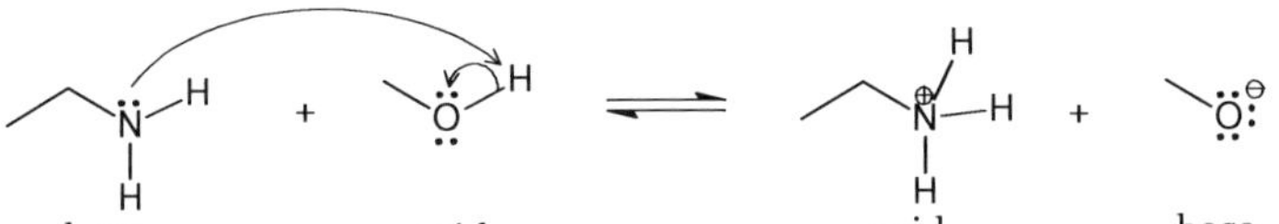

base (neutral) acid (neutral) acid (cationic) base (anionic)

Fig. 33.2

The position of equilibrium

Acid-base reactions are reversible, and products can transform back into substrates. As you know, it leads to the coexistence between substrates and products – the **state of equilibrium** is reached after some time from the start.

The important feature of every reversible chemical reaction, which has reached the state of equilibrium, is a so-called **position of equilibrium**. What is that? Simply, position of equilibrium indicates what dominates in the mixture: left side of the reaction or right side. Does the chemical system favor products or substrates? For chemists, the position of equilibrium is a measure of the efficiency. Or, to use word that is more proper – the **yield** of the reaction.

When the position of equilibrium is *shifted toward the right side* of the chemical equation, in the state of equilibrium within the flask, there are more product molecules than substrates. We can treat such reaction as efficient (yields over 50%). We denote such reactions with the double arrow pointing to the right side:

substrates $\rightleftharpoons$ products *Fig. 33.3*

When the position of equilibrium is *shifted toward the left side* of the chemical equation, in the state of equilibrium there are more substrates than products. We can treat such reaction as unfavorable (yields below 50%). We denote it with the following double arrow:

substrates $\rightleftharpoons$ products *Fig. 33.4*

Predicting the position of equilibrium with pK_a values

A universal tool to determine the position of equilibrium in acid-base reaction is **pK_a**. *It is a measure of acidity.* Unfortunately, we cannot deduce numerical values of pK_a. Chemists rather gather them by meticulous experimentation, and – bless the heavens – we do not need to know the theory to use them in practice. You will find tables with pK_a in Google, or your teacher will provide some, when needed.

Now, I will show you what we do with pK_a. In every acid-base reaction, there are two structures, which we can sign as acids, one for each side of the arrow. Look at the reaction of ethanamine and methanol: methanol is an acid on the left side, while when we read the reaction from right to left, a protonated amine plays such a role.

Then, we find values of pK_a for both these acidic components. I have used Google to find out that pK_a (CH_3OH) = 15.5, while pK_a ($CH_3CH_2NH_3^+$) = 10.6. And here comes the rule of thumb:

The lower the pK_a the stronger the acid.

Therefore, $CH_3CH_2NH_3^+$ is a stronger acid than CH_3OH. Now look once more at the chemical equation (Fig. 33.2) and treat it as two competing processes:

process "*from left to right*"
in which CH_3OH shows off its acidic properties

process "*from right to left*"
in which $CH_3CH_2NH_3^+$ shows off its acidic properties

Values of pK_a prove that $CH_3CH_2NH_3^+$ wins the challenge. Therefore, the process "from right to left" dominates, and the position of equilibrium is shifted toward substrates. We had better draw the equation with an appropriate arrow:

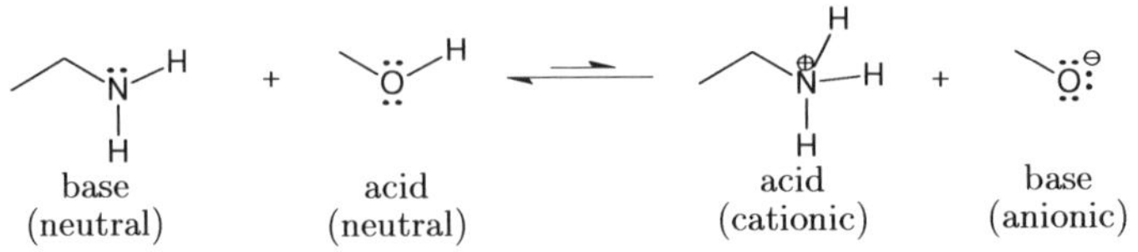

base (neutral) + acid (neutral) ⇌ acid (cationic) + base (anionic)

Fig. 33.5

In the reaction flask, there are more neutral substrates, than ionic products. The reaction is rather low yielding – when someone mixes ethanamine with methanol the reaction between the two barely takes place (only small part of substrate molecules will react to form ionic products).

It is not surprising, because although amines are rather good bases, alcohols are weak acids. So what should we expect when we mix ethanamine – a good base – with, let us say ethanoic acid – a good acid? Consider the example:

NH2 + OH ? NH3 + O

Fig. 33.6

Tables say that pK_a (CH_3COOH) = 4.8. Since pK_a ($CH_3CH_2NH_3^+$) = 10.6, CH_3COOH must be a stronger acid than $CH_3CH_2NH_3^+$. Thus the process "from left to right," where CH_3COOH shows off its acidic properties, will be the more favorable one. The position of equilibrium is in favor of the products:

NH2 + OH ⇌ NH3 + O

Fig. 33.7

In the reaction flask, there are much more ionic species, than neutral substrates. We can say that the reaction is efficient.

Predicting the position of equilibrium by deduction

The power of pK_a lies in the fact that it is a universal method, and will let us to determine the position of equilibrium in all acid-base reactions. The only condition is that you must have the access to experimental values of pKa for molecules in question.

However, there are many acid-base reactions, where we can deduce the position of equilibrium, without referring to pK_a tables. It is usually the case in acid-base reactions of the two following types:

1. a reaction with two *neutral bases* on both sides of the equation
2. a reaction with two *anionic bases* on two sides of the equation

In both situations we can simply compare the strength of two bases (we know how to do it from the previous chapter), and then the stronger base shows us the preferred direction. Consider the example of "type 1" acid-base reaction:

base (neutral) | acid (cationic) | acid (cationic) | base (neutral)

Fig. 33.8

The reaction is in favor of the products. We need to compare the strength of bases. Here, we have $CH_3CH_2NH_2$ on the left side and $PhNH_2$ on the right. Obviously, ethanamine is a stronger base, because its lone pair is not delocalized.

Therefore the process "from left to right," in which ethanamine shows off its superior basicity, will dominate in the system. No pK_a values were needed.

Now, look at the "type 2" acid-base reaction:

acid (neutral) | base (anionic) | base (anionic) | acid (neutral)

Fig. 33.9

We handle it the same way – compare the strength of two basic components, and it becomes clear, that the one on the left side is stronger. The reaction is shifted toward products, because in "left to right" process, the left anionic base shows off its superior basicity.

Problems

Sign all components of the following reactions as "acid" or "base", and deduce what is the position of equilibrium (use an appropriate double arrow), without referring to pK_a tables:

33.1

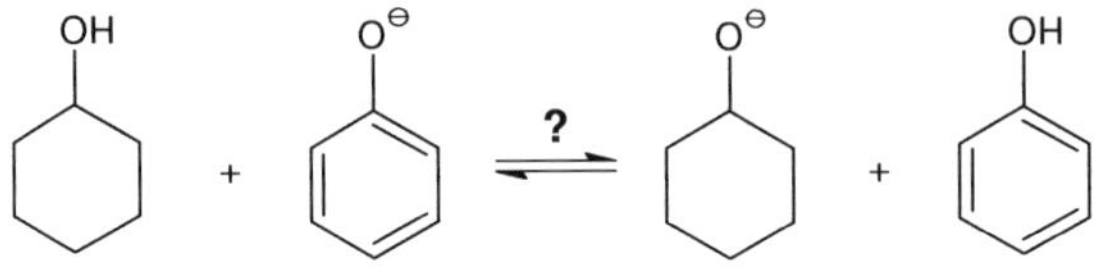
OH
O⊖
?
O⊖
OH

33.2

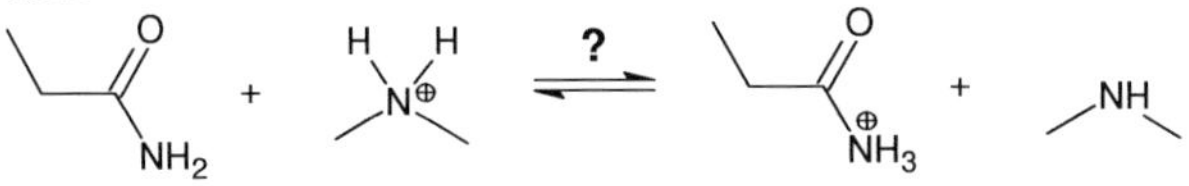
O
NH2
H H
N⊕
?
O
⊕
NH3
NH

33.3

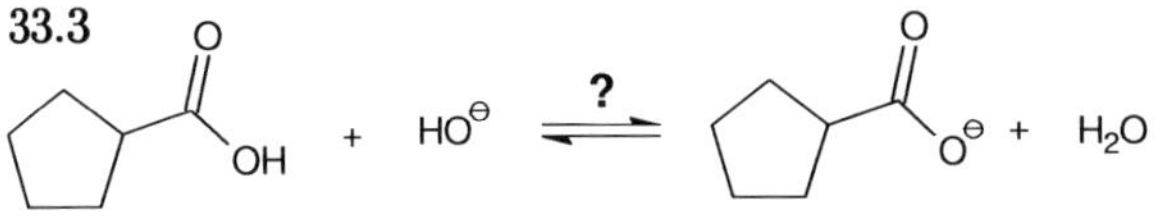
O
OH
HO⊖
?
O
O⊖
H2O

33.4

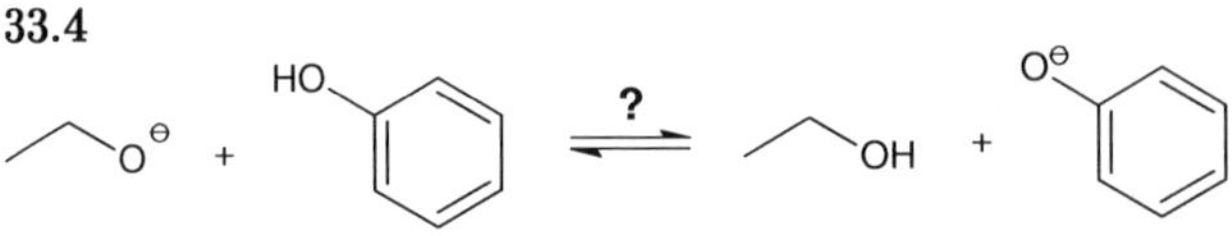
O⊖
HO
?
OH
O⊖

33.5

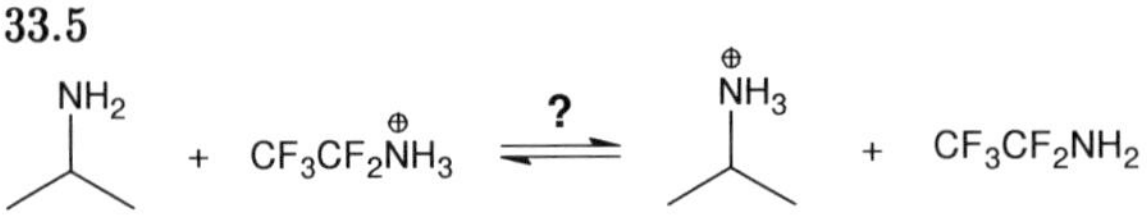
NH2
CF3CF2NH3 ⊕
?
⊕
NH3
CF3CF2NH2

33.6

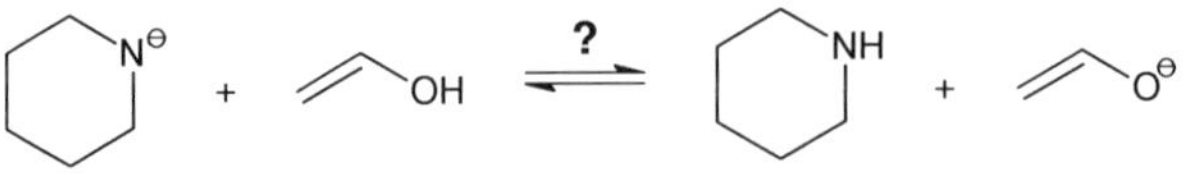
N⊖
OH
?
NH
O⊖

33.7

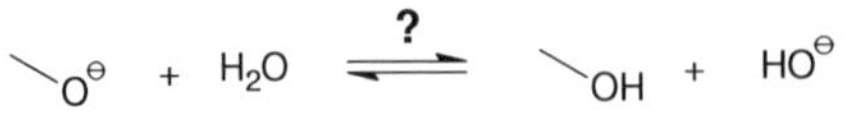
O⊖
H2O
?
OH
HO⊖

33.8

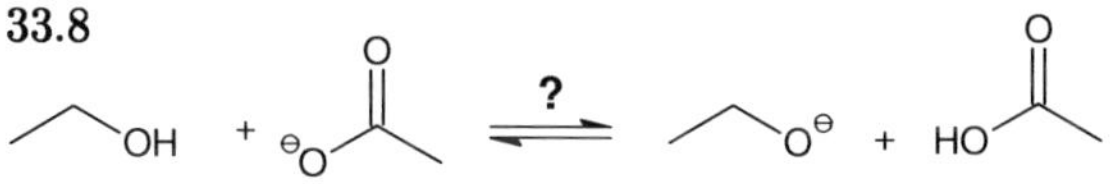

OH + $^{\ominus}$O (O) $\overset{?}{\rightleftharpoons}$ $O^{\ominus}$ + HO (O)

33.9

$O^{\ominus}$ + NH $\overset{?}{\rightleftharpoons}$ OH + $\overset{\ominus}{N}$

33.10

$H_3O^{\oplus}$ + NH $\overset{?}{\rightleftharpoons}$ H_2O + $\overset{\oplus}{N}H_2$

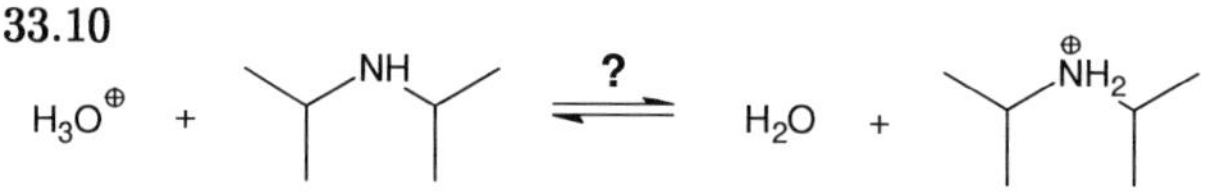

Answers to problems

Chapter 2

2.1

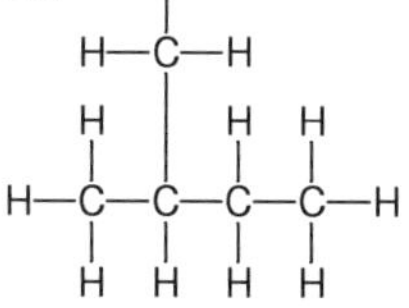

2.2

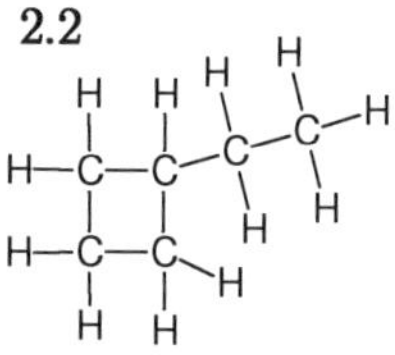

2.3

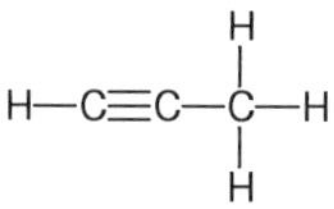

2.4

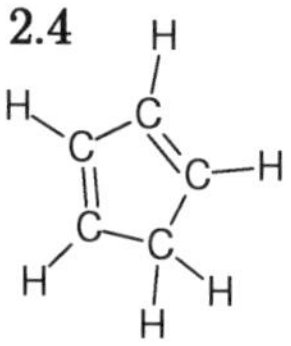

2.5

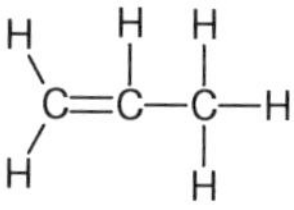

Comment: do not forget to fulfill bonding needs of all atoms. For example, forgetting about H atom at one of C≡C bond in molecule **2.3** is a mistake.

2.6

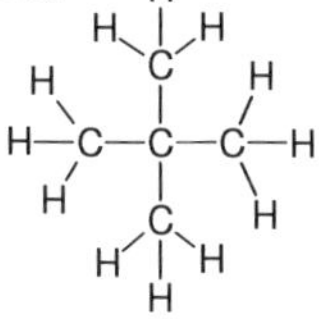

2.7

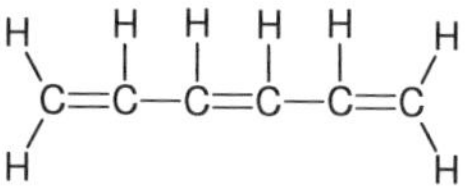

2.8

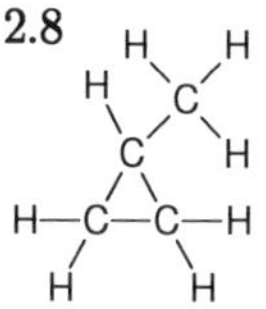

2.9

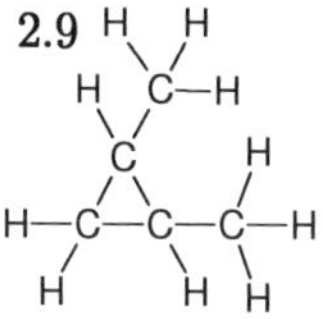

2.10

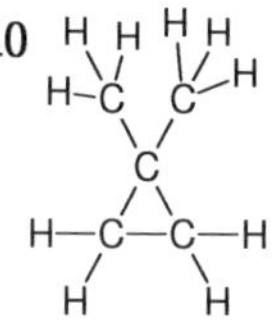

Comment: Note that in hydrocarbons **2.6** and **2.10** there are C atoms with no H atoms attached – all of their four bonds lead to other Cs. We call such C atoms **quaternary carbon atoms**.

2.11 There are two pairs of isomeric molecules: **2.1** with **2.6** (C_5H_{12}), and **2.9** with **2.10** (C_5H_{10}).

Chapter 3

3.1 and

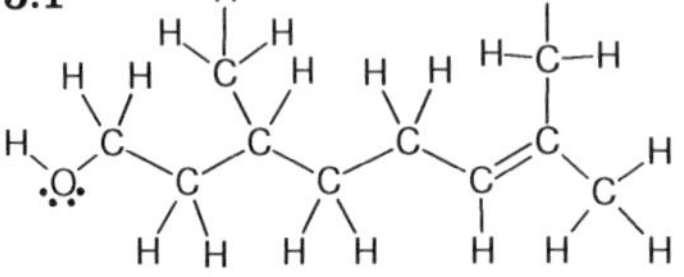

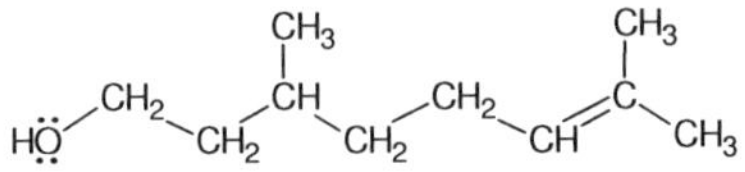

3.2 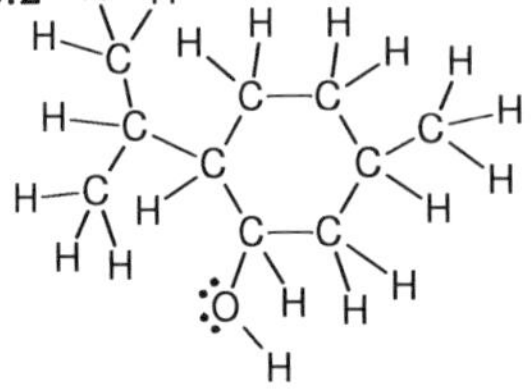 and

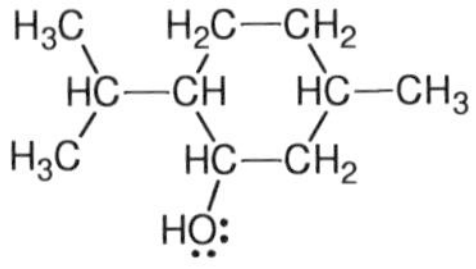

3.3

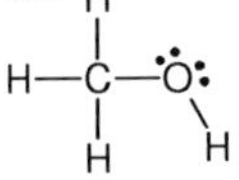

3.4

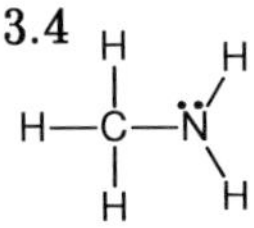

3.5

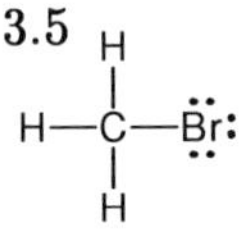

3.6

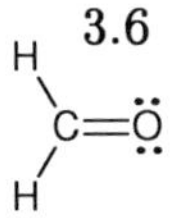

3.7

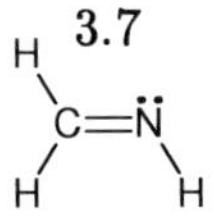

3.8

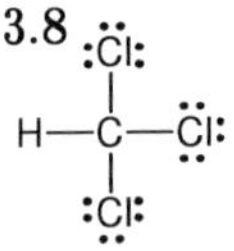

3.9

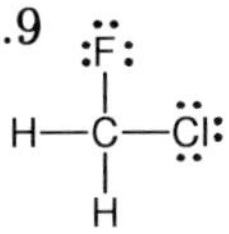

3.10

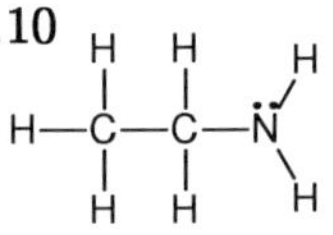

3.11

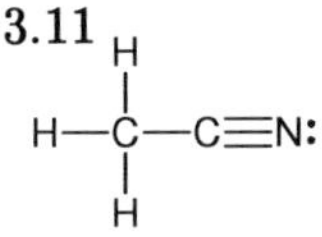

3.12

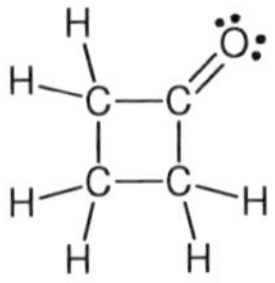

3.13

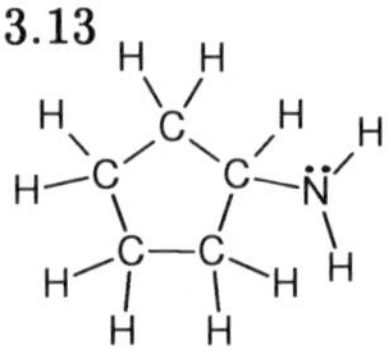

3.14

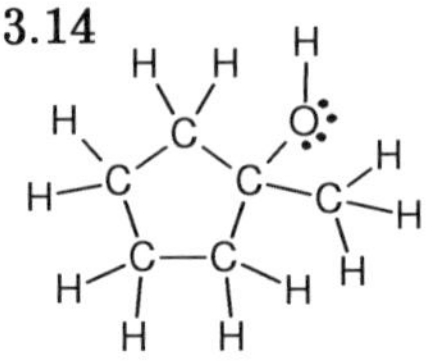

3.15 There are no isomers among molecules **3.3-3.14**.

3.16 and

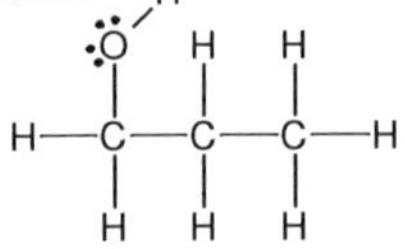

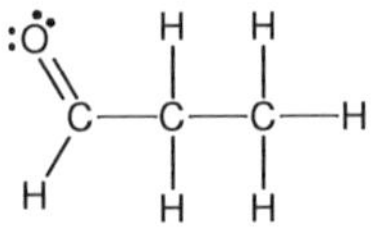

3.17 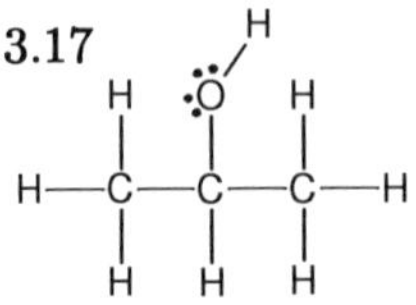 and

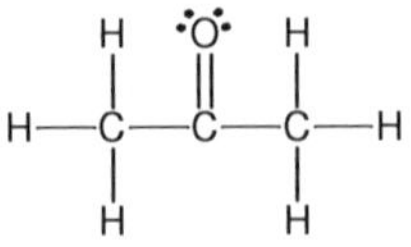

3.18 :Br—C—C—N (with H atoms) and :Br—C—C=N—H and :Br—C—C≡N:

3.19 Since we can rotate, flip and flop drawings, it would be the same molecule as the one in problem **3.16**. We can even say that "there is no third C." Both terminal Cs are identical, equivalent and "first."

3.20

3.21

3.22

3.23

3.24

3.25

Chapter 4

4.1-4.9

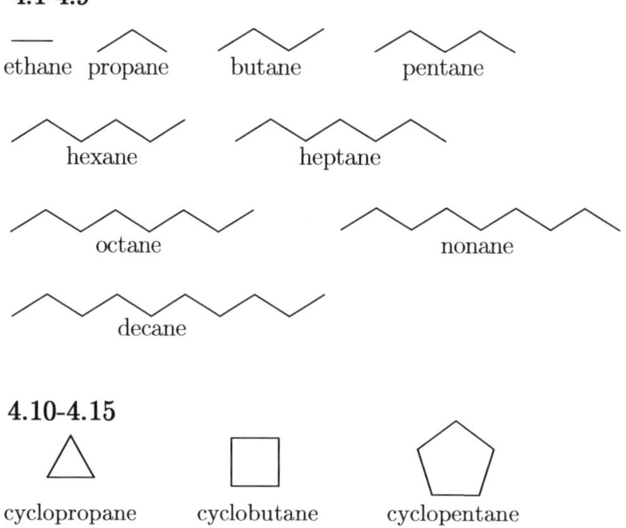

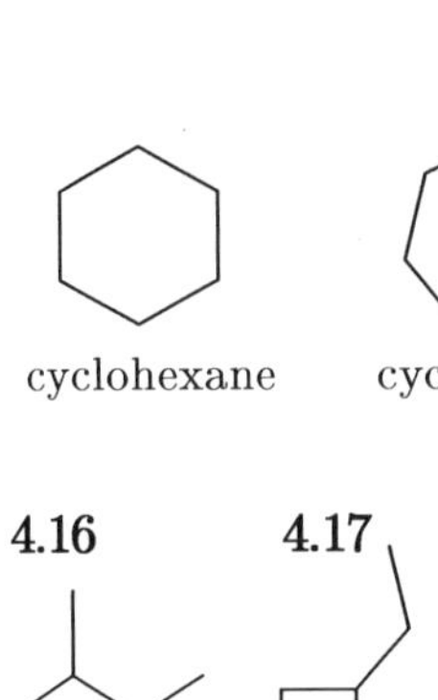

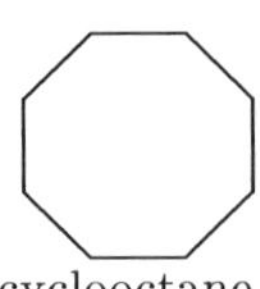

cyclohexane cycloheptane cyclooctane

4.16 4.17 4.18 4.19 4.20 4.21

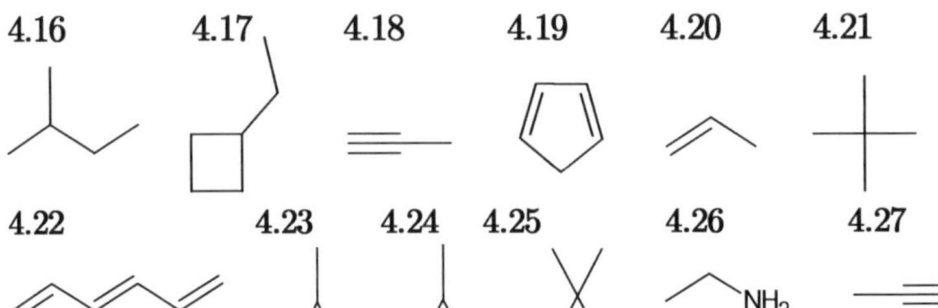

4.22 4.23 4.24 4.25 4.26 4.27

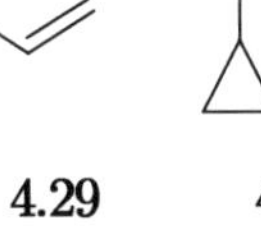

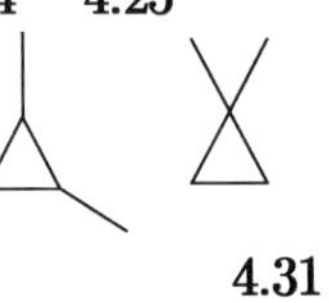

4.28 4.29 4.30 4.31 4.32

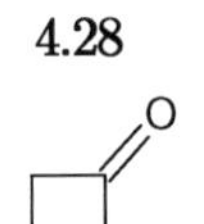

4.33 4.34 4.35 4.36 4.37

4.38 4.39 4.40 4.41 4.42 4.43

4.44 4.45 4.46 4.47

4.48 4.49 4.50 4.51

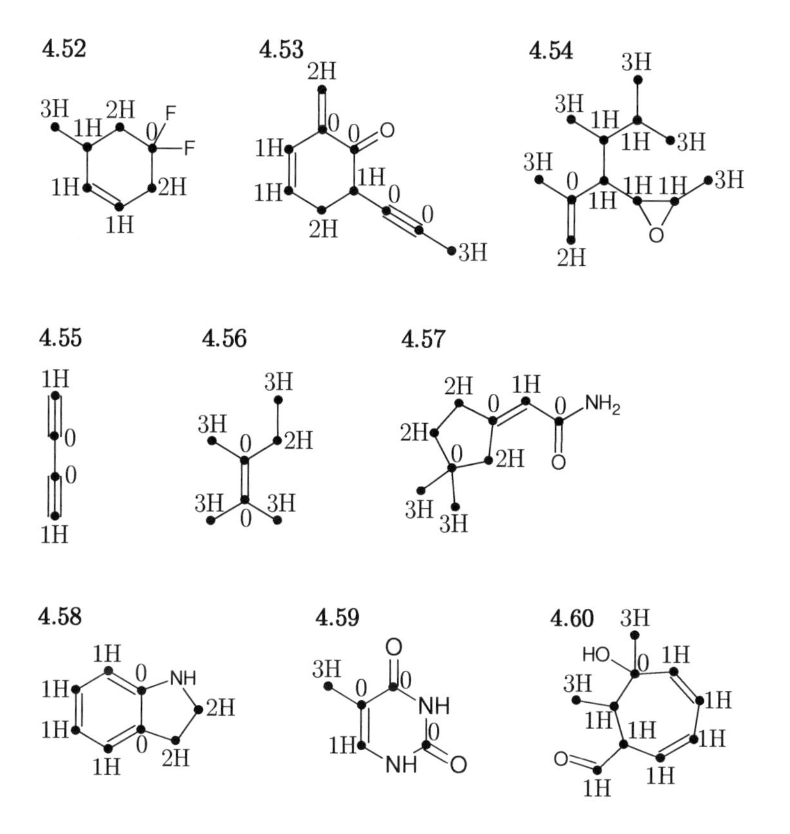
4.52
4.53
4.54
4.55
4.56
4.57
4.58
4.59
4.60

Chapter 6

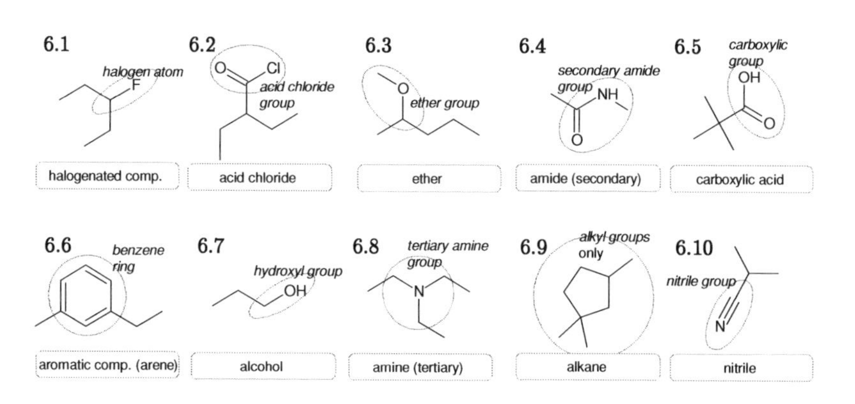
6.1
halogen atom
halogenated comp.
6.2
acid chloride group
acid chloride
6.3
ether group
ether
6.4
secondary amide group
amide (secondary)
6.5
carboxylic group
carboxylic acid
6.6
benzene ring
aromatic comp. (arene)
6.7
hydroxyl group
alcohol
6.8
tertiary amine group
amine (tertiary)
6.9
alkyl groups only
alkane
6.10
nitrile group
nitrile

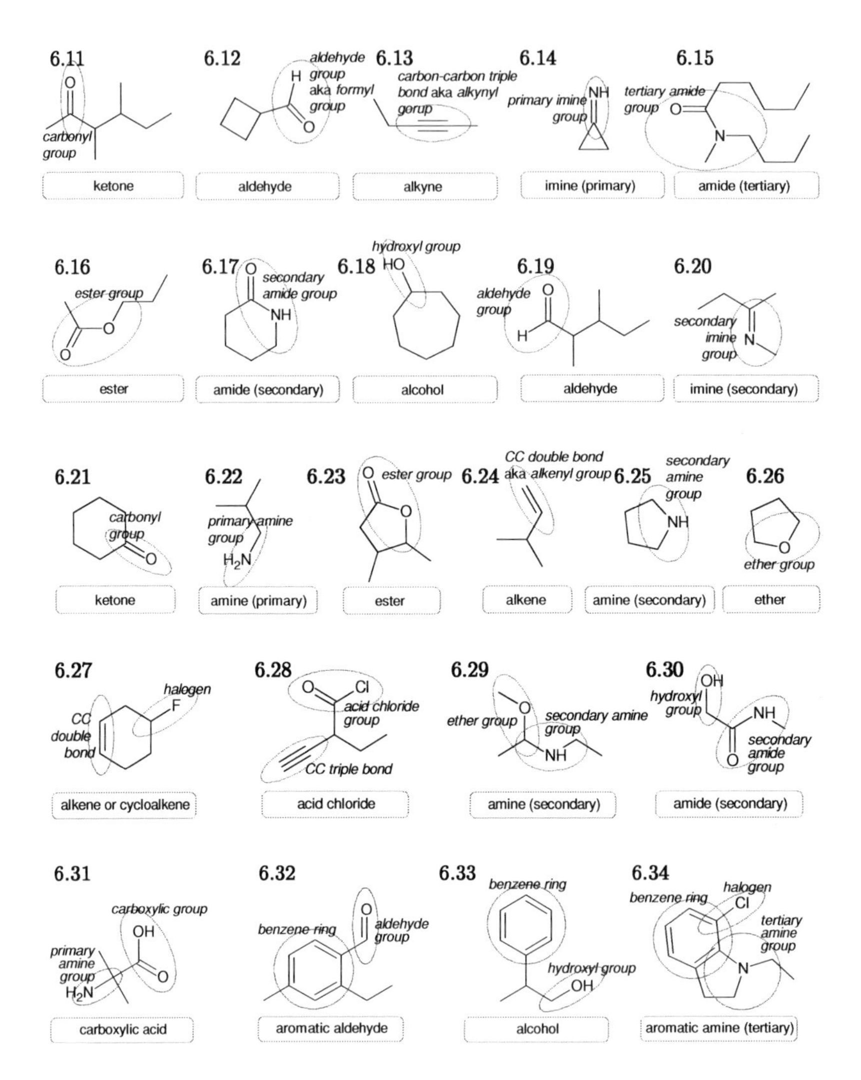

Comment on 6.33: Note that this molecule is not classified as aromatic alcohol, because the hydroxyl group does not bond to the benzene ring directly.

6.35 CC double bond, CC triple bond, halogen — alkene (cycloalkene)

6.36 carbonyl group, nitrile group — nitrile

6.37 carbonyl group, aldehyde group, carbonyl group — aldehyde

6.38 hydroxyl group, aldehyde group — aldehyde

6.39

ester

6.40

ester

6.41

amine (secondary)

6.42

ester

6.43

amide (secondary)

6.44

ketone

6.45

aldehyde

6.46

alcohol

6.47

carboxylic acid

6.48

diamine

6.49

amide

6.50

ketone

6.51

carboxylic acid chloride

6.52

alcohol

6.53

amine

6.54

ketone

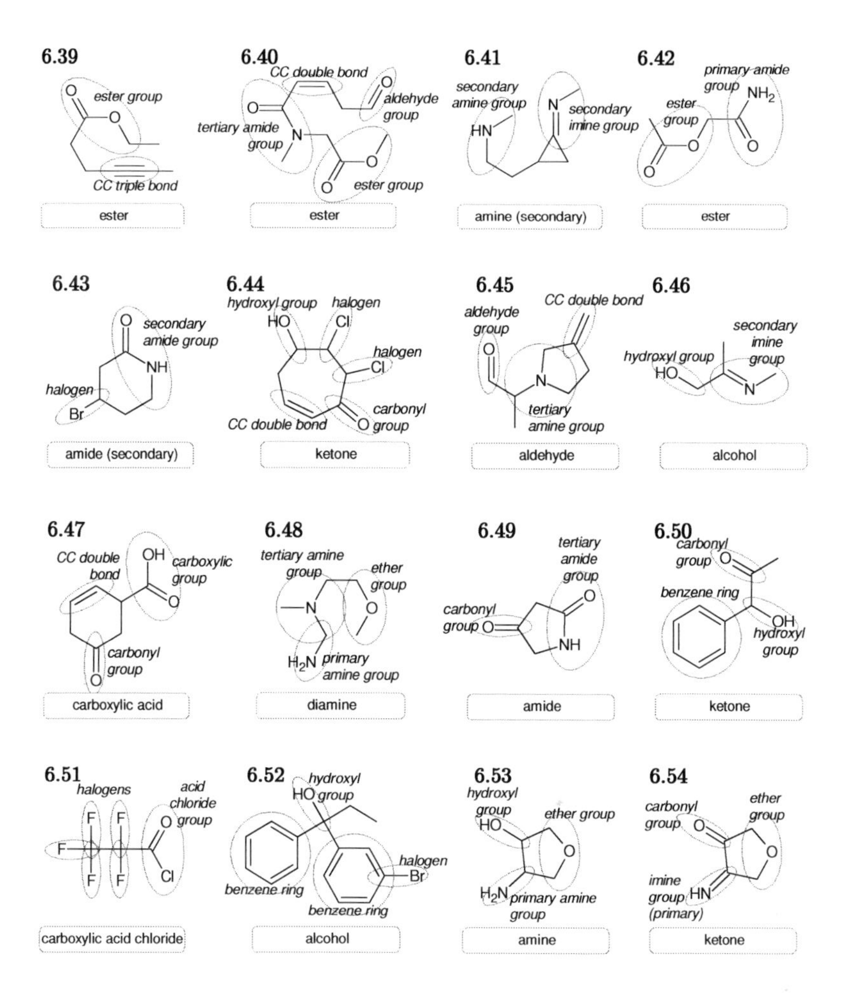

6.55 It is an example of an organic molecule, which contains no H atoms (formula: C_3ClF_5O).

6.56

diketone

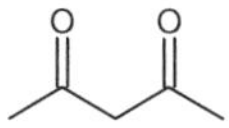

6.57

diamine

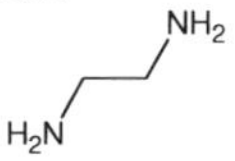

6.58

pentaalkene

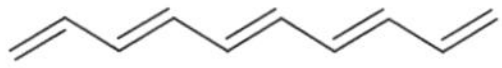

6.59

hexaalcohol

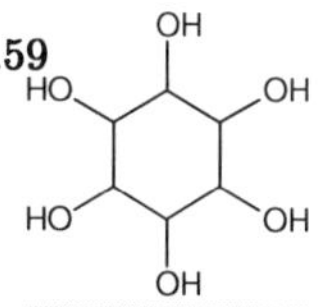

6.60

tetraaldehyde

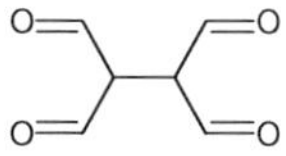

6.61

trialcohol

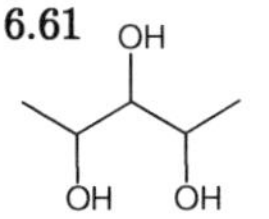

6.62

dialkyne

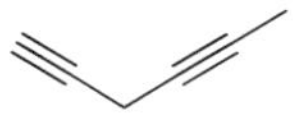

6.63

dicarboxylic acid

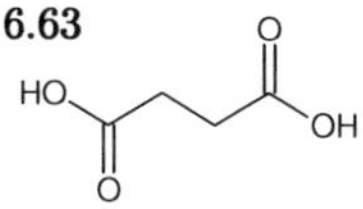

Comment: Chemists often refer to alcohols with more than one hydroxyl group as diols, triols, tetraols, instead of dialcohols, trialcohols, tetraalcohols, etc.

Chapter 7

7.1	**ol** chunk for an OH group and **fluoro** for a F atom
7.2	**trione** chunk for three carbonyl groups
7.3	**al** chunk for CHO and **dione** for two carbonyl groups
7.4	**oic acid** chunk for COOH and **imino** for C=NH
7.5	**amine** chunk for NH_2, **en** for C=C, and **yn** for C≡C
7.6	**amide** chunk for $CONH_2$, **hydroxy** for OH, and **phenyl** for a benzene ring
7.7	**benzene** for a benzene ring, and **dibromo** for Br atoms
7.8	**nitrile** for a C≡N group, **formyl** for CHO, and **en** for C=C
7.9	**oyl chloride** for COCl, **oxo** for carbonyl and yn for C≡C

7.10 *root*

7.11 1 2 *root* 3 4 5 6

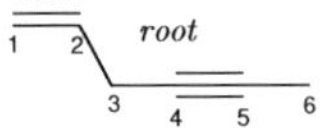

7.12 *root*

7.13 *root*

7.14 1... *root*

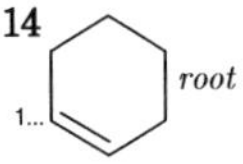

7.15 $-\!\!\xi\!\!-CH_3$ or $-\!\!\xi\!\!-Me$ (1)
the smallest (1 C) *branch* possible

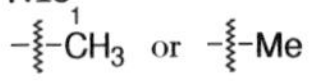

7.16 1...

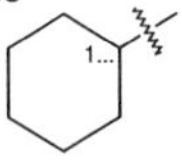

7.17 1 2 3 4 5... *root*

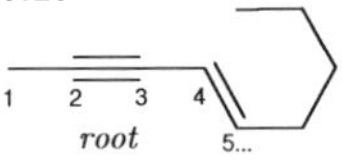

7.18 1... *root*

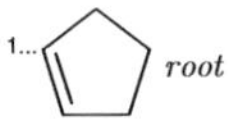

7.19 *root*

7.20 1 2 3...

7.21 1 2 or $-\!\!\xi\!\!-Et$

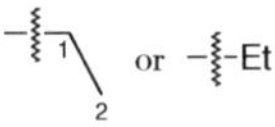

7.22 1 2 3 4 5 6 *root*

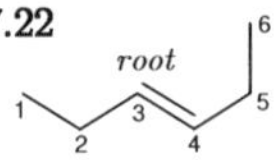

7.23 1 2...

7.24 1 2... *root*

7.25 1 2... *root*

7.26 1 2 3... *root*

7.27 1 2...

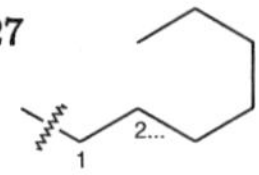

7.28 1 2 3 4 5... *root*

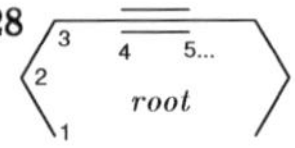

7.29 1...

7.30 1 2...

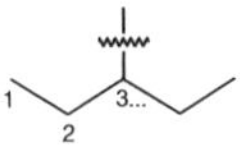

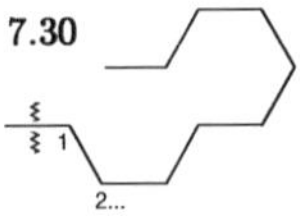

7.31 1...

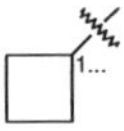

7.32 C *root*

7.33 *root*

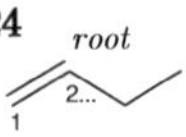

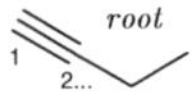

7.34 1 2 3... *root*

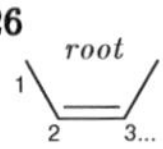

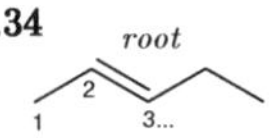

7.35 1 2... *root*

7.36 *root*

7.37

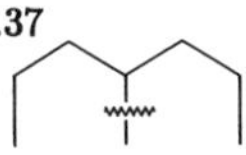

7.38 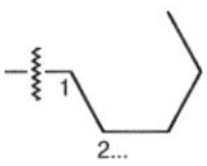 1 2...

7.39 1 2 3 4 *root*

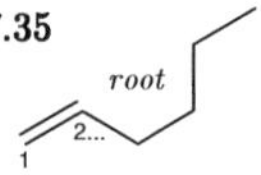

7.40 1... *root*

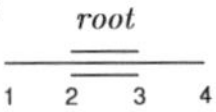

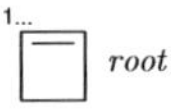

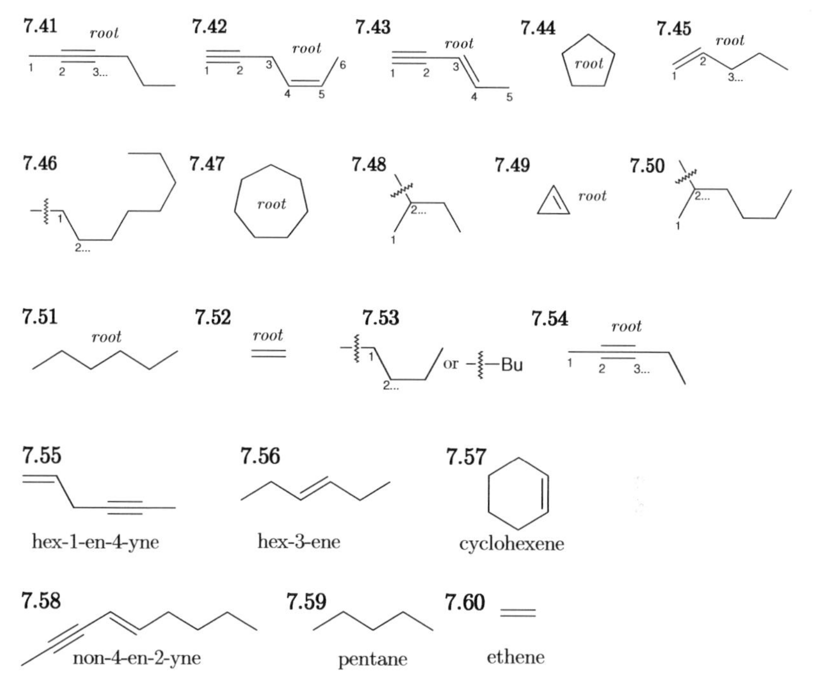
7.41
root
1
2
3...
7.42
root
1
2
3
4
5
6
7.43
root
1
2
3
4
5
7.44
root
7.45
root
1
2
3...
7.46
1
2...
7.47
root
7.48
1
2...
7.49
root
7.50
1
2...
7.51
root
7.52
root
7.53
1
2...
or
Bu
7.54
root
1
2
3...
7.55
hex-1-en-4-yne
7.56
hex-3-ene
7.57
cyclohexene
7.58
non-4-en-2-yne
7.59
pentane
7.60
ethene

Chapter 8

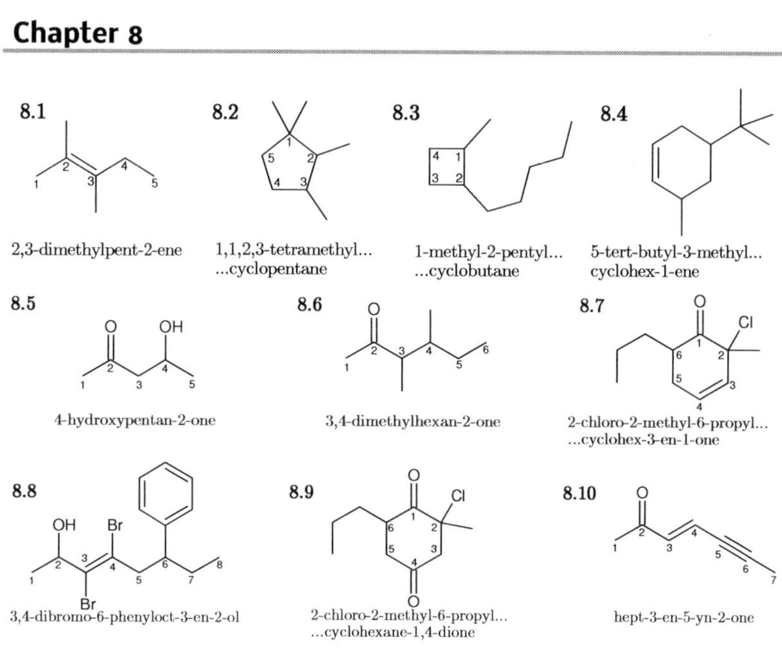
8.1
2,3-dimethylpent-2-ene
8.2
1,1,2,3-tetramethyl...
...cyclopentane
8.3
1-methyl-2-pentyl...
...cyclobutane
8.4
5-tert-butyl-3-methyl...
cyclohex-1-ene
8.5
O
OH
4-hydroxypentan-2-one
8.6
O
3,4-dimethylhexan-2-one
8.7
O
Cl
2-chloro-2-methyl-6-propyl...
...cyclohex-3-en-1-one
8.8
OH
Br
Br
3,4-dibromo-6-phenyloct-3-en-2-ol
8.9
O
Cl
O
2-chloro-2-methyl-6-propyl...
...cyclohexane-1,4-dione
8.10
O
hept-3-en-5-yn-2-one

8.11

2-phenyl-5-propylcyclopentan-1-imine

8.12

2-amino-3,3-diphenylpropan-1-ol

8.13

2-sec-butylcyclopent-3-en-1-amine

8.14

1,2-dimethylcyclo...
...hepta-4,6-diene-1,3-diol

8.15

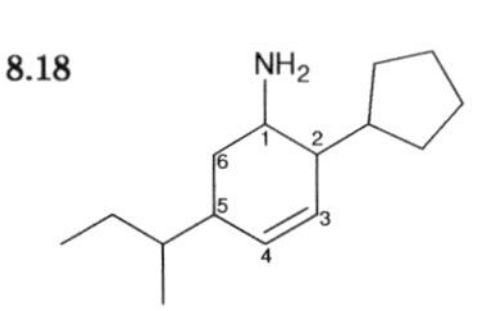

3-ethyl-4-imino-6-methyloctan-2-ol

8.16

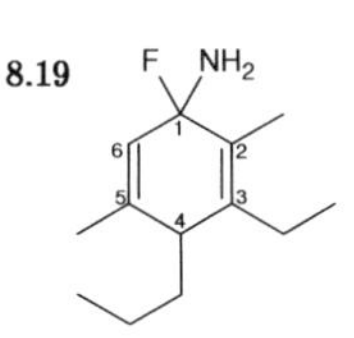

1-cyclopentylpentan-2-ol

8.17

2-cyclopropyl-3-ethyloct-5-en-4-one

8.18

5-(butan-2-yl)-2-cyclopentyl...
...cyclohex-3-en-1-amine

8.19

3-ethyl-1-fluoro-2,5-dimethyl...
...-4-propylcyclohexa-2,5-dien-1-amine

8.20

Chapter 9

9.1 but-3-yn-2-one (ketone)
9.2 2-bromobutan-1-ol (alcohol)
9.3 2-bromopropane-1,3-diol (dialcohol a.k.a. diol)

Comment on 9.3: you should add "**e**" to the root's chunk, when the first letter in the highest priority functional group chunk *is a consonant.* Here: "diol," so "propan" becomes "propan**e**."

9.4 2,4,5-trifluoropent-1-ene (alkene)
9.5 4-methylpent-1-en-2-ol (alcohol)
9.6 2-cyclopentylethan-1-amine (primary amine)
9.7 1-fluorobut-3-en-2-imine (primary imine)
9.8 3-ethylpentan-2-one (ketone)
9.9 4-ethyl-3-methylheptane (alkane)
9.10 3,4-dimethylhexane (*not* 2-ethyl-3-methylpentane!) (alkane)
9.11 4,5-diethylhept-1-ene (alkene)

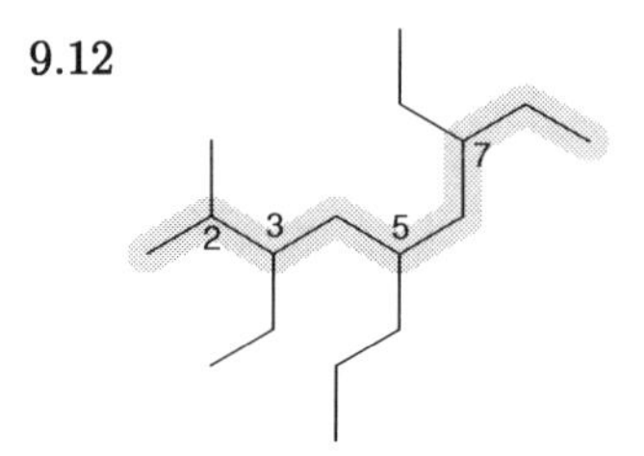

3,7-diethyl-2-methyl-5-propylnonane

9.13-17 Five isomeric alkanes with C_6H_{14} formula are as follows:

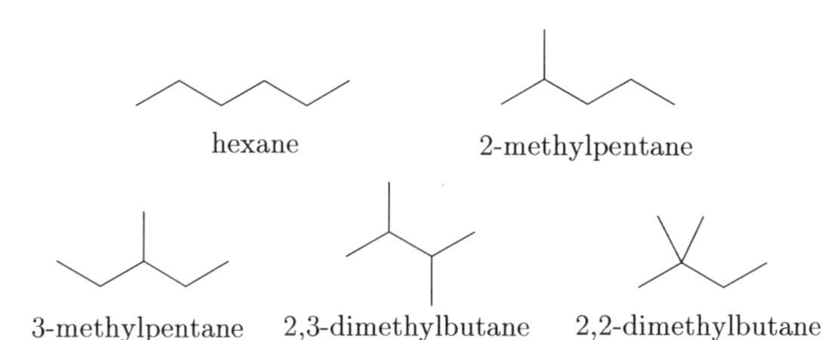

9.18 6-methylheptan-3-one
9.19 2-fluoro-5-methylhex-2-en-3-ol
9.20 3,5-difluorohex-5-ene-2,4-diol

Comment on 9.20: letter "**e**" is inside the root's chunk, because there is *a consonant* afterwards (**d** in **d**iol chunk)

9.21 6-bromo-5-hydroxyoctan-3-one
9.22 6-bromo-5-hydroxyoct-5-en-1-yn-3-one
9.23 4,5,6-triaminohexan-2-one
9.24 3-methyl-1-phenyl-2-en-4-yn-1-amine
9.25 4,5-dibromo-1-cyclopentyl-1-hydroxydec-4-en-2-one
9.26 2-methylheptan-4-one
9.27 2,2,4-trimethylpentan-3-one
9.28 2-hydroxy-4-methylpentan-3-one
9.29 cyclopentanamine (*not* cyclopentan-1-amine!)
9.30 2-aminocyclohexan-1-ol
9.31 cyclopent-2-en-1-amine
9.32 4-methylcyclopent-2-en-1-ol
9.33 3-methylcyclopent-3-en-1-ol
9.34 cyclooct-4-en-1-one
9.35 3-hexylcyclobutan-1-one
9.36 3-ethyl-3-methylcyclobutan-1-one
9.37 cyclohexanimine (*not* cyclohexan-1-imine!)
9.38 1-phenylcyclopropan-1-amine

9.39 1-methyl-2-(propan-2-yl)cyclohexane

Comment on 9.39: Another acceptable name is 2-isopropyl-1-methylcyclohexane (recall that isopropyl is a popular name for (propan-2-yl) branch; it goes before methyl, because the letter "i" precedes "m" in the alphabet).

9.40 2,4-dimethylcycloheptane-1,3-dione
9.41 3-butylcycloprop-1-ene
9.42 2-phenylcyclohexan-1-ol
9.43 cyclohexylbenzene

Comment on 9.43: Did you fell into a trap? This molecule is *not* phenylcyclohexane, because the benzene ring is the root – it wins in priority. Cyclohexyl plays the role of a branch. Moreover, correct IUPAC name is cyclohexylbenzene, *not* 1-cyclohexylbenzene, because locant "1" can be omitted, causing no ambiguities.

9.44-55 Twelve structurally isomeric cycloalkanes with C_6H_{12} formula are as follows:

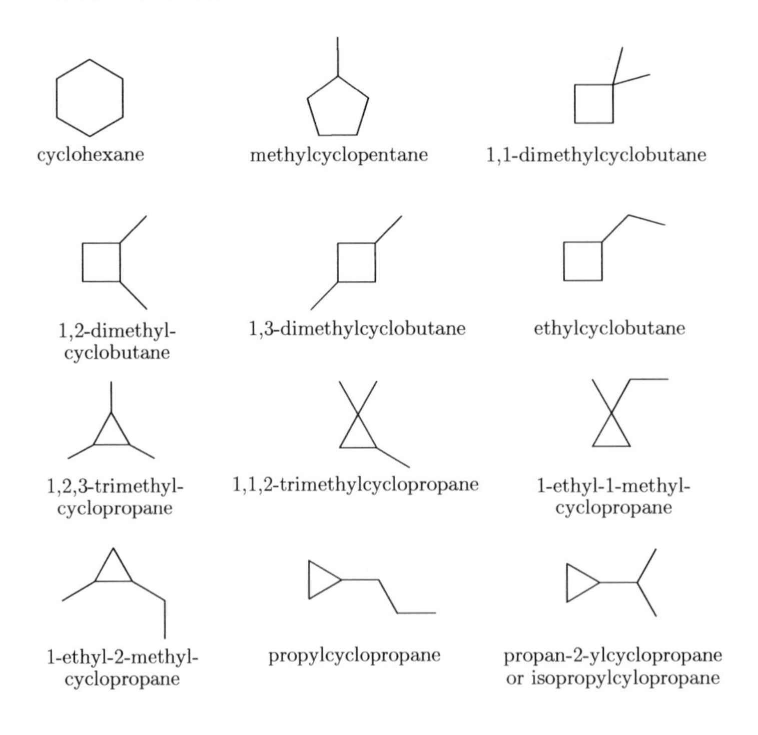

9.56 3-*tert*-butylhexan-2-one
9.57 3-ethyl-4,4-dimethylpentan-2-one

Comment on 9.57: you had better follow IUPAC rules and not use a *tert*-butyl, unless unnecessary. It was necessary in **9.56**, though, because the hexan root was clearly different from the branched branch. However, in **9.57** we can spot a less obvious pentan root, so that there is no hassle with a *tert*-butyl. Name 3-ethyl-4,4-dimethylpentan-2-one is therefore better than alternative 3-*tert*-butylpentan-2-one, because IUPAC recommends keeping *branches as simple as possible*. Take a look:

better choice of the root
(recommended by IUPAC)

acceptable choice,
but not recommended

9.58

9.59

9.60

9.61 5-hydroxyhex-3-ynal
9.62 2,3-dimethyl-4-oxobutanenitrile
9.63 2-ethyl-3-fluorobutanoyl chloride
9.64 2-ethyl-3-fluorobutanamide
9.65 2,4-dimethylhex-3-en-5-ynenitrile
9.66 2-cyclohexylethanal

9.67

9.68

9.69

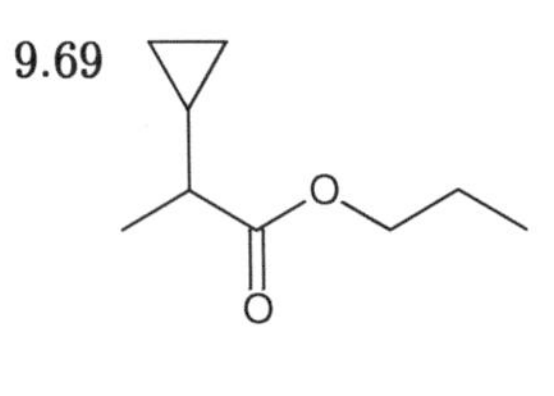

9.70 ethyl 2-fluoropropanoate
9.71 methyl 2-ethylbut-3-enoate
9.72 propan-2-yl 3-cyclopentylpropanoate

Chapter 10

10.1

10.2

10.3

10.4

10.5

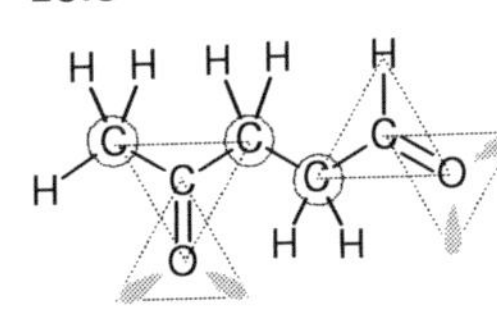

10.6

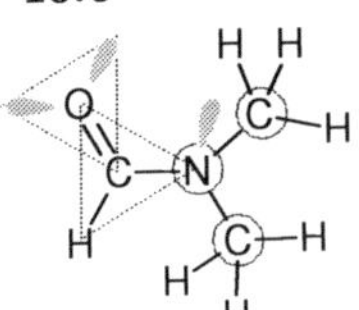

10.7

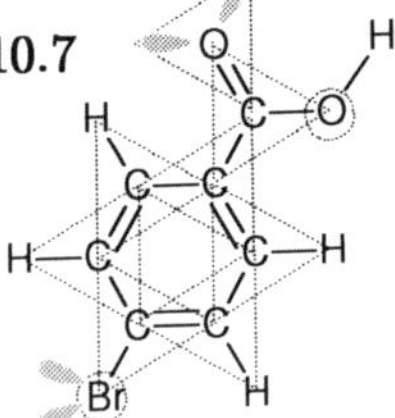

10.8

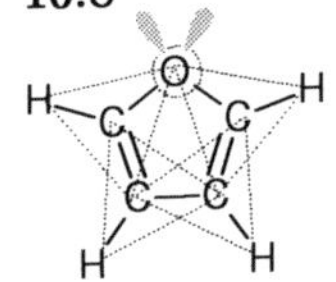

10.9

10.10

Chapter 11

11.1

staggered
(the most stable)

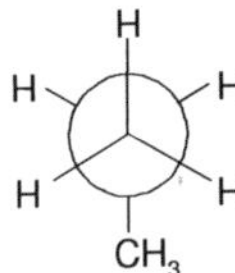

eclipsed
(the least stable)

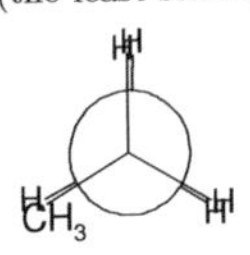

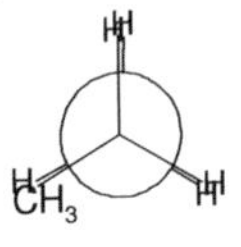

11.2

staggered
(the most stable)

eclipsed
(the least stable)

11.3

staggered
(the most stable)

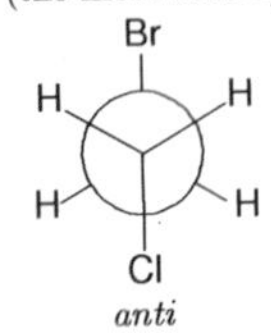

anti

eclipsed

staggered

gauche

eclipsed
(the least stable)

11.4

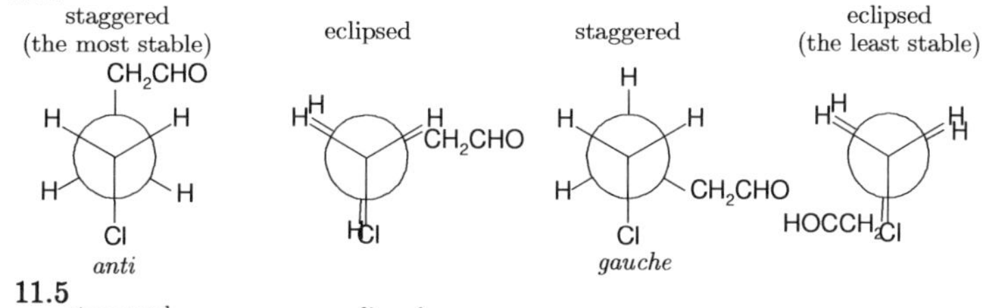

staggered (the most stable) — eclipsed — staggered — eclipsed (the least stable)

CH_2CHO, H, H, H, H, Cl — *anti*

H, H, H, CH_2CHO, H, Cl

H, H, H, H, CH_2CHO, Cl — *gauche*

H, H, H, H, $HOCCH_2$, Cl

11.5

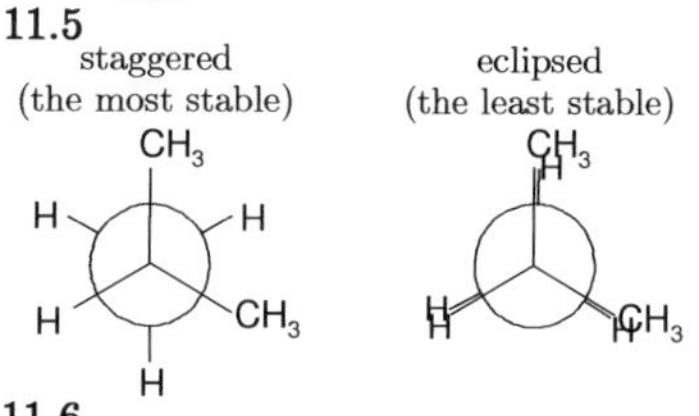

staggered (the most stable) — eclipsed (the least stable)

CH_3, H, H, H, CH_3, H

CH_3, H, H, H, H, CH_3

11.6

staggered — eclipsed (the least stable) — staggered (the most stable) — eclipsed

CH_3, H, CH_3, H, CH_3, H

CH_3, H, H, H, CH_3, CH_3

CH_3, H, H, H, CH_3, CH_3

CH_3, H, H_3C, H, H, CH_3

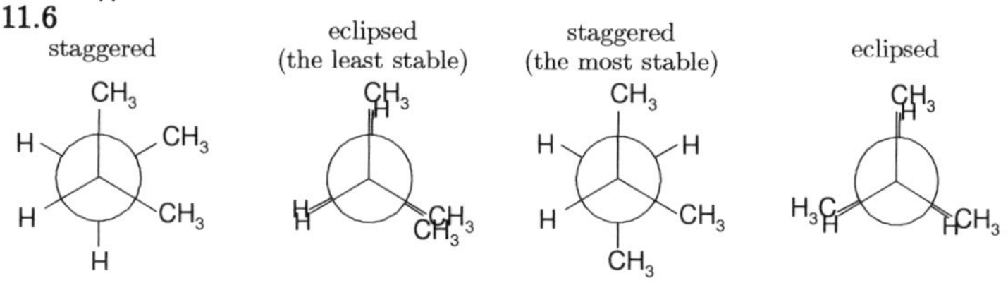

11.7 B (both front and back groups have been rotated; those two pictures are two different conformations of hexane)

11.8 A (the second picture can be perceived as flipped version of the first)

11.9 C (note that Br and CH_3 bond frontal C, while in the other molecule to the back C)

11.10 D

11.11 A

11.12 B

11.13 E (it is hexane in both projections; first one is drawn along the C3-C4 bond, while the second along C2-C3)

11.14 C

Chapter 14

14.1

14.2

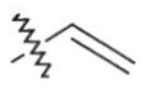

14.3 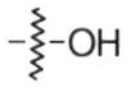 -OH

14.4 Ph

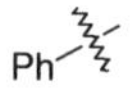

14.5 Ph

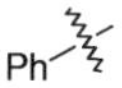

14.6

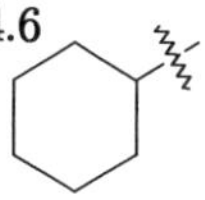

14.7 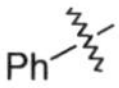 Ph

14.8 Cl

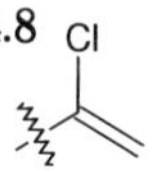

14.9 **14.10** 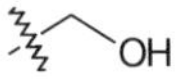 **14.11** 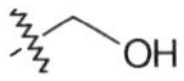 **14.12**

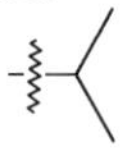

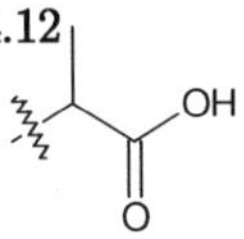

OH OH OH O

14.13 –Br, $–NCH_3$, $–CH_2Cl$, $–CH_2F$, $–CH_2NH_2$, $–CH_2Ph$
14.14 $–NHCH_3$, $–CH_2Cl$, $–CHFCH_3$, $–CH_2F$, $–CH_2OH$,
$–CH_2CH_2F$, $–CH_2Ph$
14.15 $–OCH_3$, $–NH_2$, $–CONH_2$, $–COCH_3$, $–CH(CH_3)_2$, $–CH_3$
14.16 $–COOH$, $–COCH_3$, $–C{\equiv}CH$, $–C(CH_3)_3$, $–CH{=}CH_2$, $–CH(CH_3)_2$
14.17 $–OC(CH_3)_3$, $–OCH_3$, $–CH_2OH$, $–C{\equiv}N$,
$–CH_2CH_2OH$, $–CH_2C(CH_3)_3$

14.18 *Z*
14.19 *Z*
14.20 *E*
14.21 *E*
14.22 *E*
14.23 *Z*

Following examples are easy cases, and they do not require drawing tree representations of substituents:

14.24 *Z* (CH_3 wins H, F wins H; CH_3 & F are on the same side)
14.25 *E* (Cl wins CH_3, OCH_3 wins H; opposite sides)
14.26 *E* (Br wins H, Cl wins F; opposite sides)
14.27 *E* (CH_2Cl wins H, CHO wins H; opposite sides)
14.28 no *E*/*Z* isomerism
14.29 *Z* (CH_2OH wins CH_3, CH_3 wins H; same side)

14.30-35 Six isomeric alkenes with a formula C_5H_{10} are as follows:

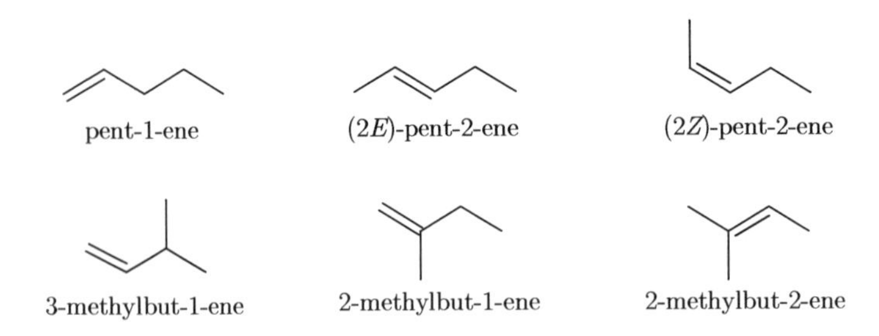

Chapter 15

15.1

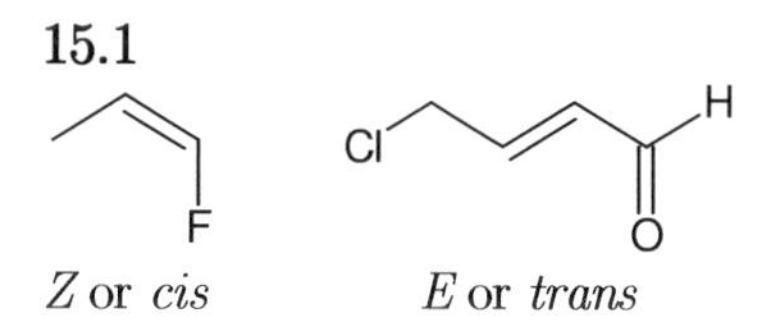

Z or *cis* *E* or *trans*

15.2-15.8 Seven isomeric dimethylcyclohexanes are as follows:

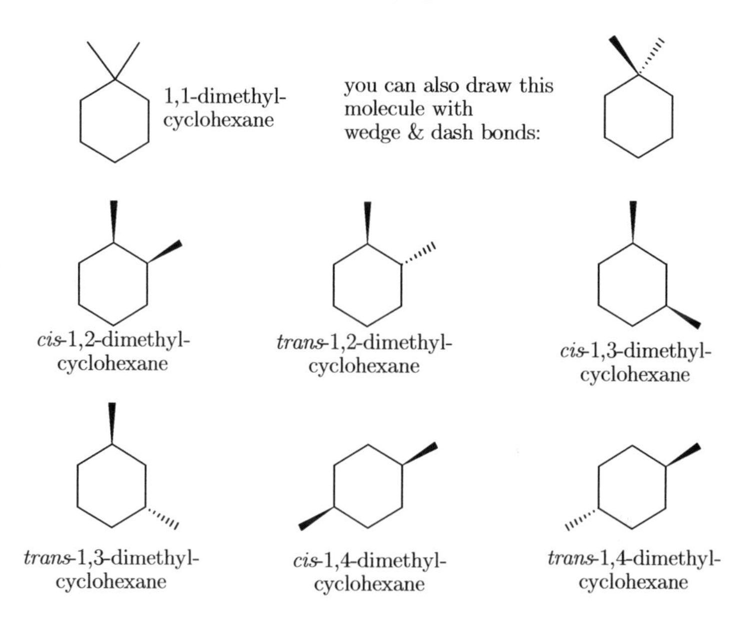

cis-1,2-dimethyl-cyclohexane *trans*-1,2-dimethyl-cyclohexane *cis*-1,3-dimethyl-cyclohexane

trans-1,3-dimethyl-cyclohexane *cis*-1,4-dimethyl-cyclohexane *trans*-1,4-dimethyl-cyclohexane

15.9 *Trans* stereoisomer of 4-methylcylohexan-1-ol is more stable. It spends nearly all of the time in chair conformation, which has both substituents located in equatorial positions. *Cis* stereoisomer is a less stable molecule, because all the time, there is one substituent attached to the ring via an axial bond.

Chapter 16

16.1 O Cl *

16.2 Cl * OH

16.3 O * *

16.4 no C*

16.5 * N

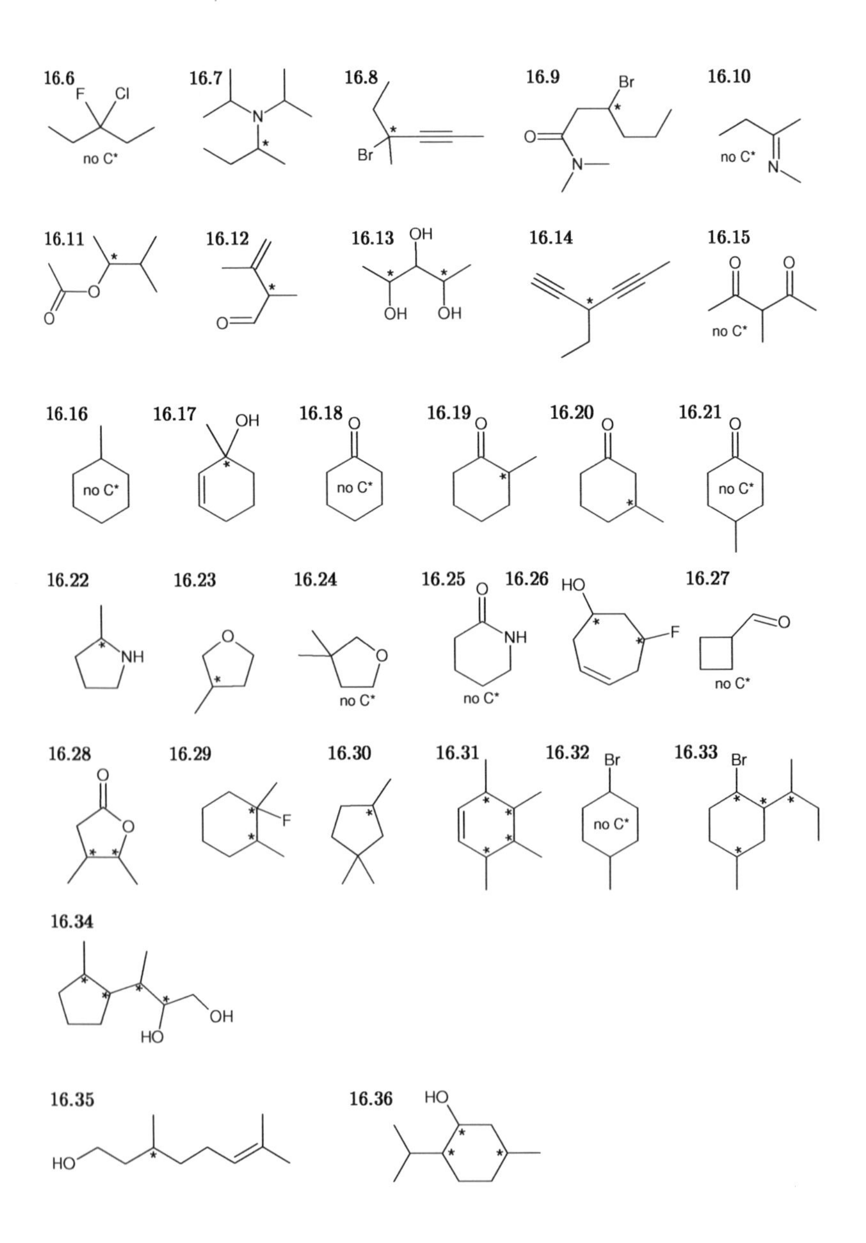

16.37-44 8 different fluoroalkanes with $C_5H_{11}F$ formula are as follows:

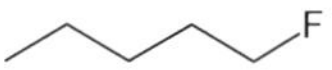

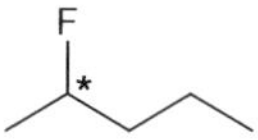

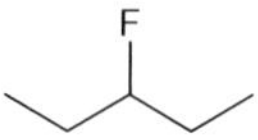

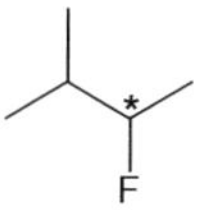

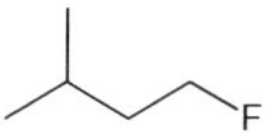

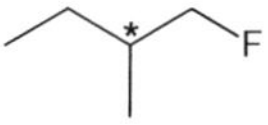

Chapter 17

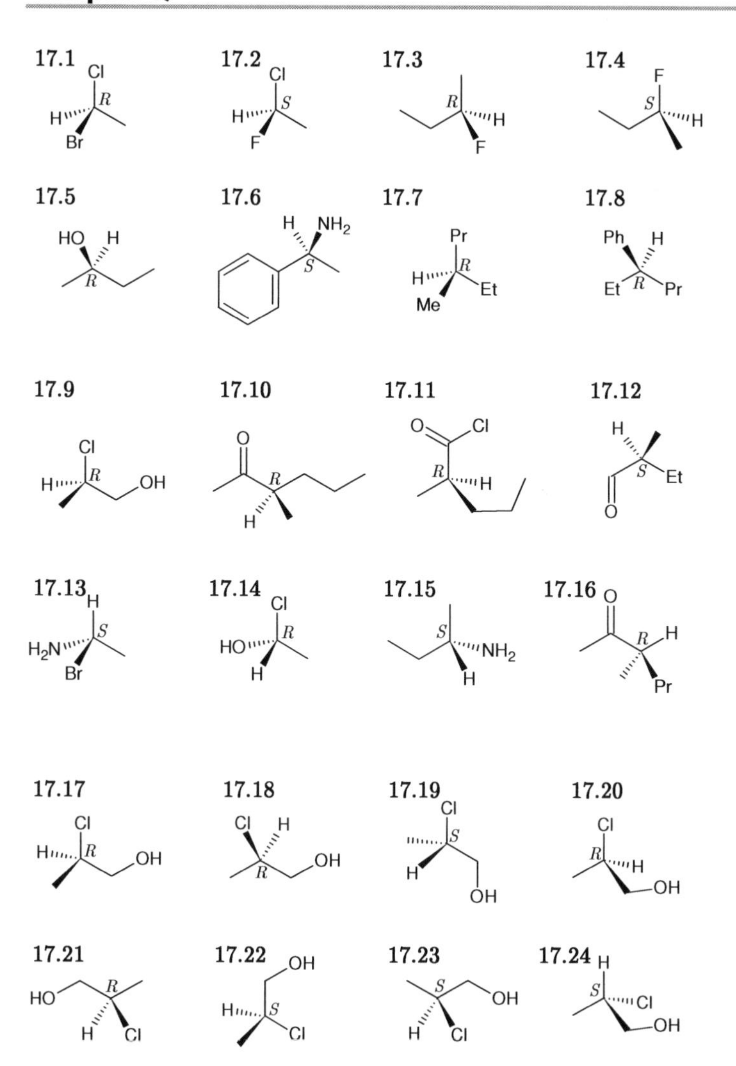
17.1
17.2
17.3
17.4
17.5
17.6
17.7
17.8
17.9
17.10
17.11
17.12
17.13
17.14
17.15
17.16
17.17
17.18
17.19
17.20
17.21
17.22
17.23
17.24

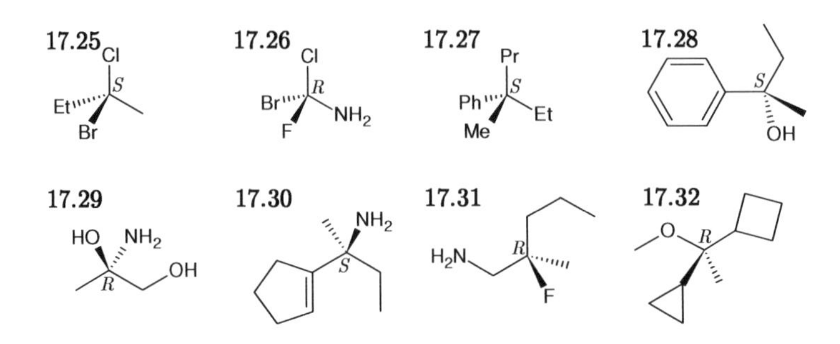

17.33 (2*R*)-2-fluoro-2-methylpentan-1-amine
17.34 (2*R*)-2-aminopropane-1,2-diol
17.35 (2*S*)-2-phenylbutan-2-ol
17.36 (3*R*)-3-methylhexan-2-one
17.37 (1*S*)-1-bromoethan-1-amine
17.38 (2*S*)-2-methylbutanal
17.39 (2*R*)-2-methylpentanoyl chloride
17.40 (3*R*)-3-methylhexane
17.41 (1*S*)-1-phenylethan-1-amine

17.42 OH *R*

17.43 *S*

17.44 HN *R* F

17.45 *R* N

17.46 Cl *R* O

17.47 NH_2 *R* COOH

17.48 *R* F

17.49 OH Et *S* O

17.50

HO

(3*S*)-3,7-dimethyloct-6-en-1-ol

HO

(3*R*)-3,7-dimethyloct-6-en-1-ol

17.51 *S*

17.52

OH *R* *R* O NH_2

(2*R*,4*R*)-4-hydroxy-2-methylhexanamide

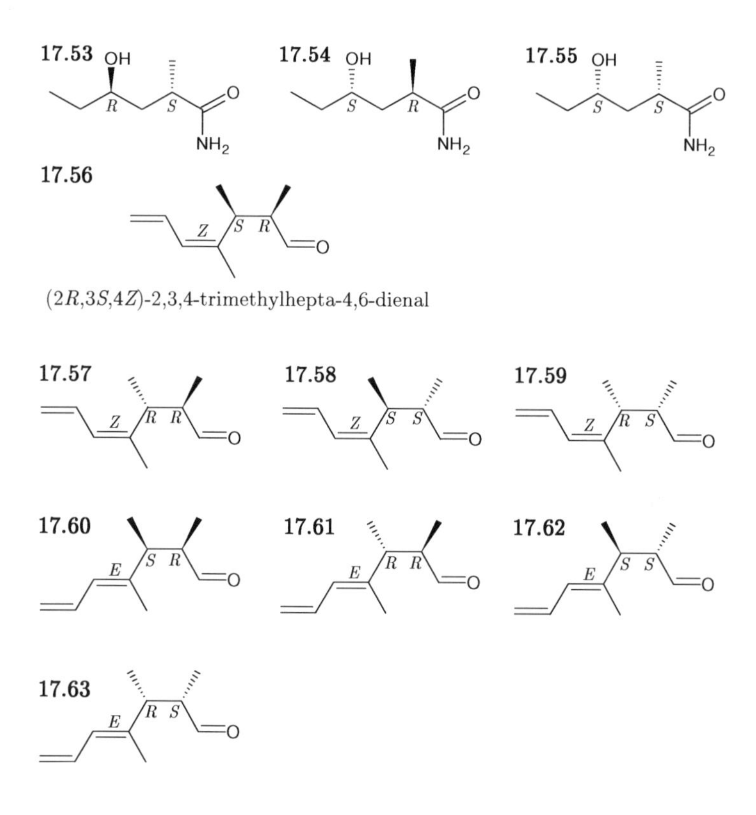

(2*R*,3*S*,4*Z*)-2,3,4-trimethylhepta-4,6-dienal

17.64 There are total of 8 different stereoisomeric menthols:

(1*R*,2*R*,5*R*)	(1*S*,2*S*,5*S*)	(1*S*,2*R*,5*R*)	(1*R*,2*S*,5*R*)
(1*R*,2*R*,5*S*)	(1*R*,2*S*,5*S*)	(1*S*,2*R*,5*S*)	(1*S*,2*S*,5*R*)

17.65 OH *R* O

17.66 OH *S* *S*

17.67 OH *S*

17.68 *S* *R*

17.69 HN *S* F H

17.70 O *S* *R*

17.71 OH *R* NH2 *R* HO

17.72 Cl Cl *R* NH2

Comment on 17.72: note, that C atom in CCl_2 is not C*. Nevertheless, there is no reason for not to use wedge and dash bonds.

17.73 *Cis/trans* are applicable only to disubstituted rings:
molecule **17.66** is *trans*
molecule **17.70** is *cis*

17.74-77

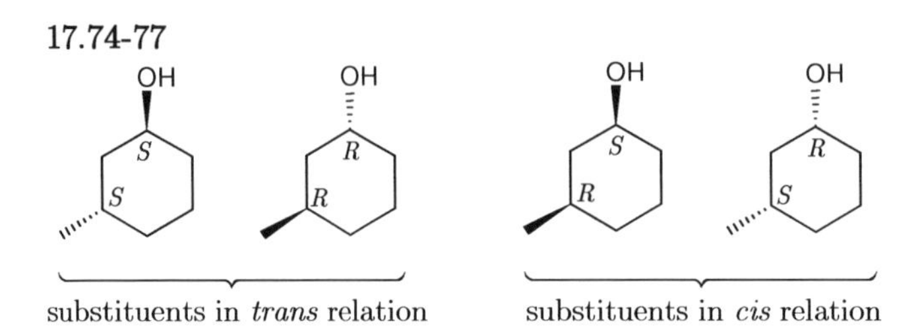

Counter-intuitively, the name *trans*-3-methylcyclohexan-1-ol describes two stereoisomers: (1*S*,3*S*)-3-methylcyclohexan-1-ol, and (1*R*,3*R*)-3-methylcyclohexan-1-ol. The name *cis*-3-methylcylohexan-1-ol describes both (1*S*,3*R*)-, as well as (1*R*,3*S*)-3-methylcyclohexan-1-ol.

Chapter 18

18.1 In menthol there are 3 stereocenters (3 C*), so $n = 3$. The total number of stereoisomers is $2^3 = 2 \cdot 2 \cdot 2 = 8$.
18.2 Stereoisomeric menthol molecules are as follows:

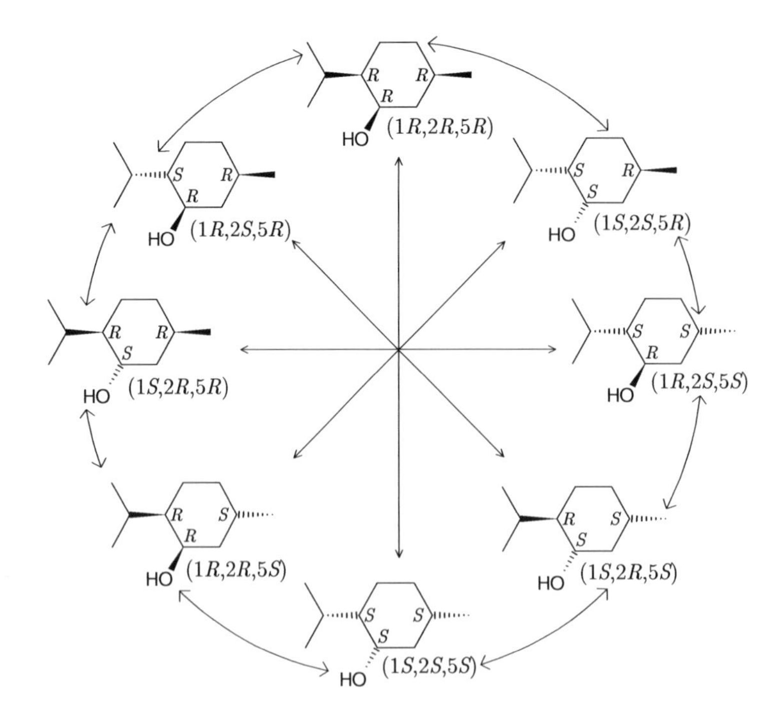

Each of these molecules is chiral and optically active.
Straight lines inside the circle point out all four enantiomeric pairs.
Curved lines on the perimeter point out examples of diastereoisomeric pairs (all other possible connections are diastereoisomeric too; and there are many more of them).

18.3

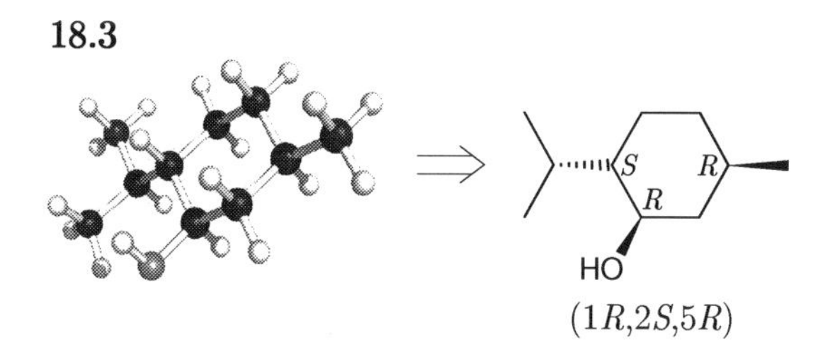

(1*R*,2*S*,5*R*)

Chapter 20

20.1

(& 3 σ pairs of C-H bonds)

20.2

(& 8 σ pairs of C-H bonds)

20.3

(& 7 σ pairs of C-H bonds)

20.4

(& 4 σ pairs of C-H bonds)

20.5

(& 9 σ pairs of C-H bonds)

Chapter 21

21.1 delocalized lp & π pair

21.2

21.3 delocalized lp & π pairs

21.4 delocalized π pairs

21.5 delocalized lp & π pairs

21.6 delocalized π pairs

21.7 delocalized π pairs

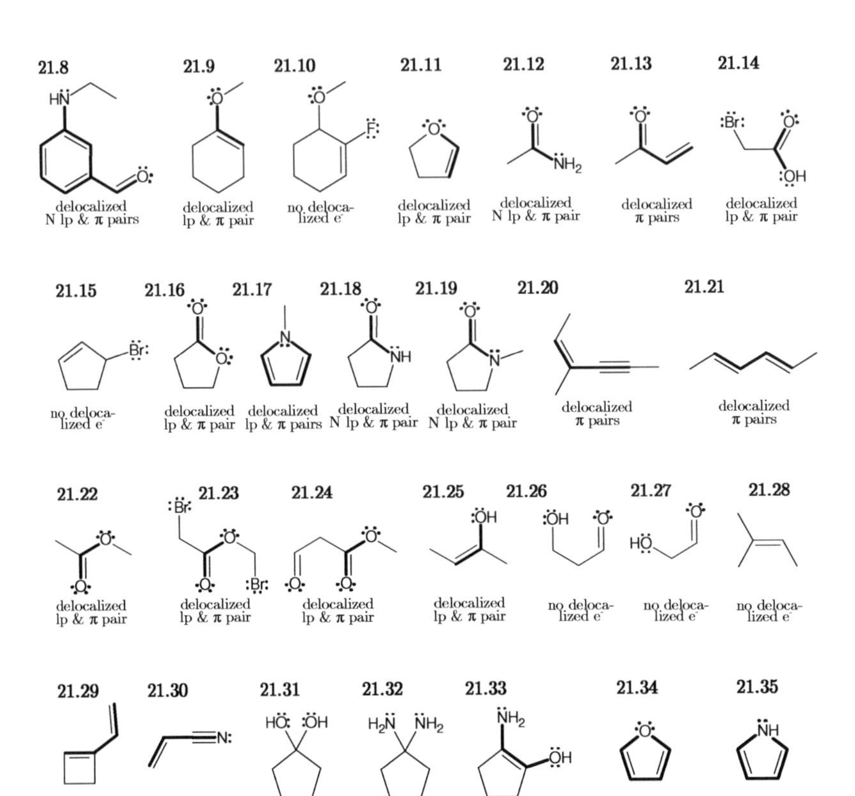

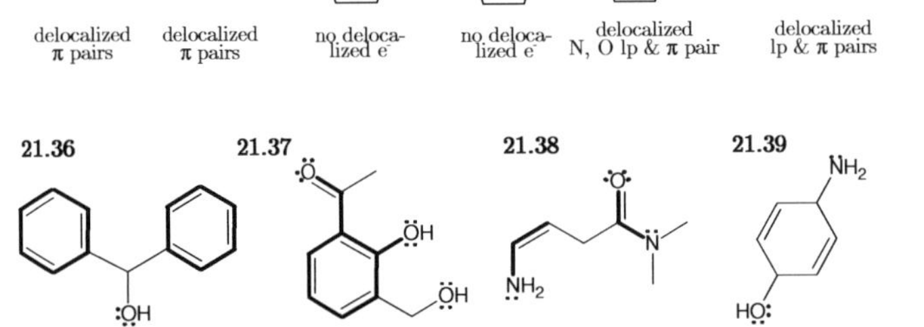

Chapter 22

22.1

22.2

22.3

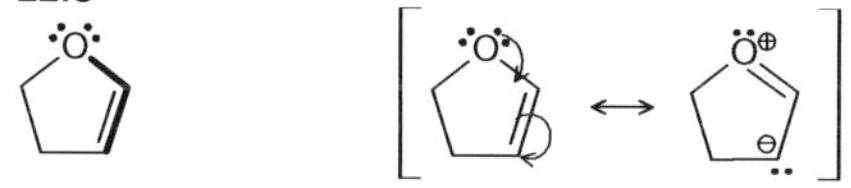

22.4

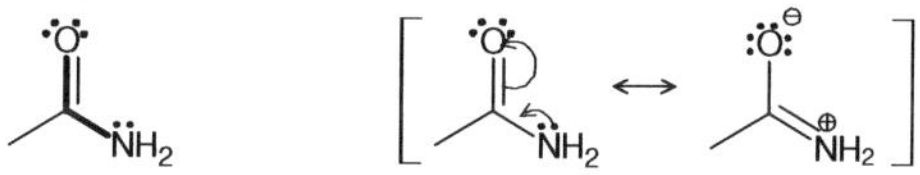

22.5

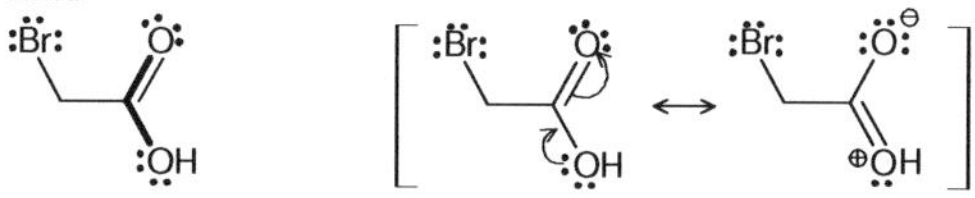

22.6

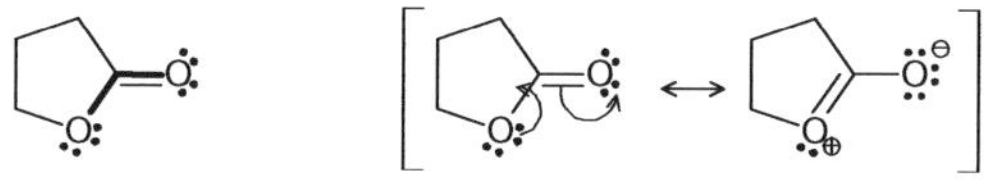

22.7

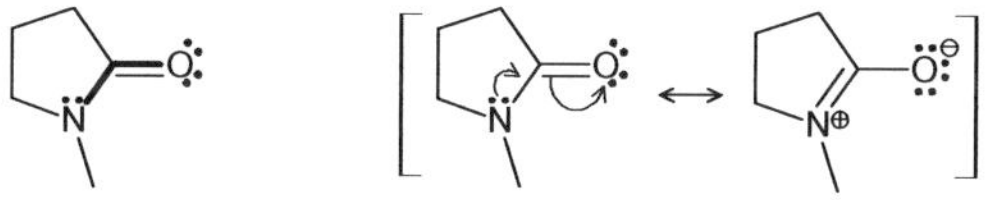

22.8

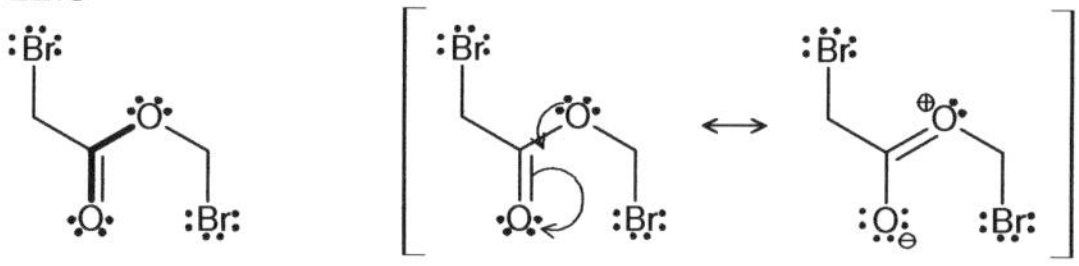

22.9

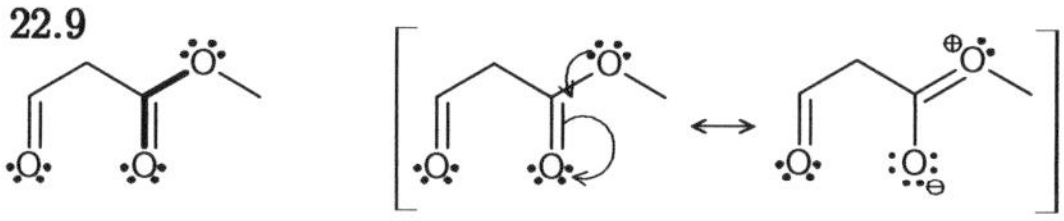

22.10

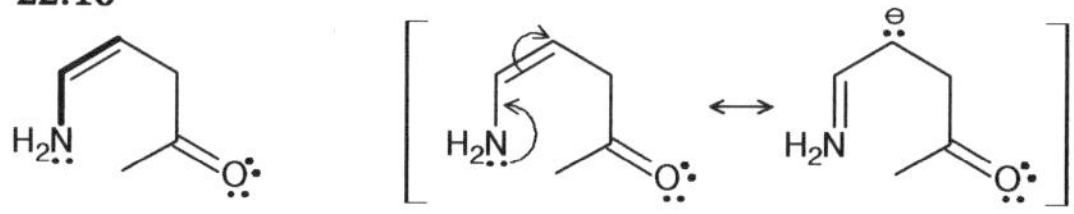

22.11

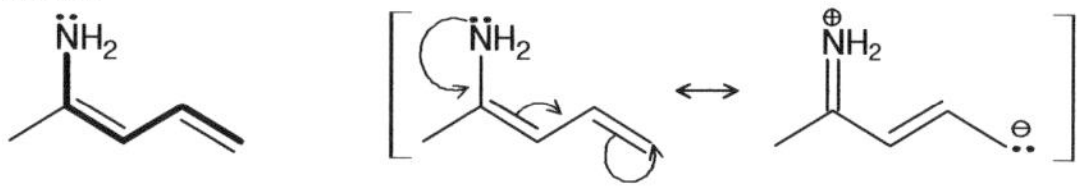

22.12

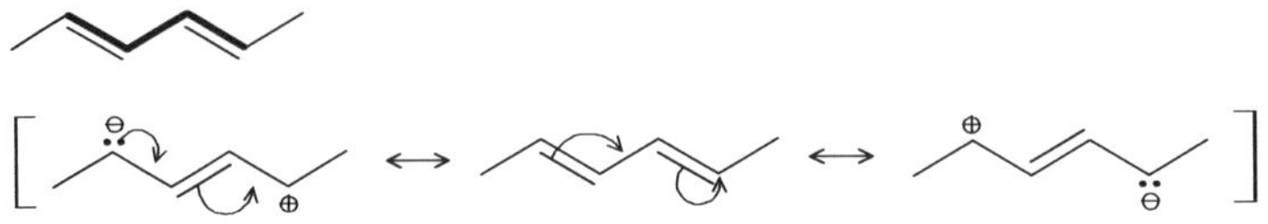

Comment on 22.12: formally, left and right structures are two identical drawings, just flopped. Nevertheless, we must draw both of them, and we say that the molecule has three resonance structures.

22.13

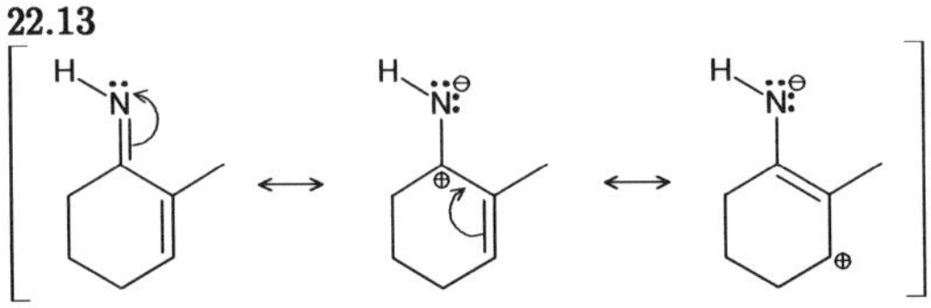

22.14

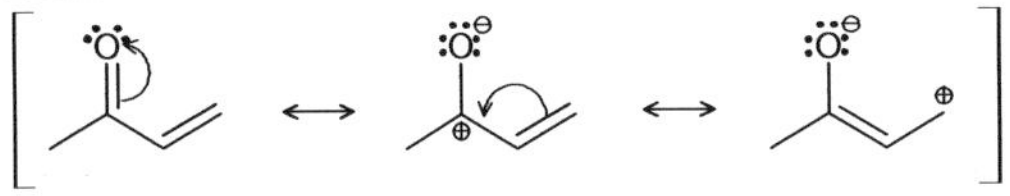

22.15

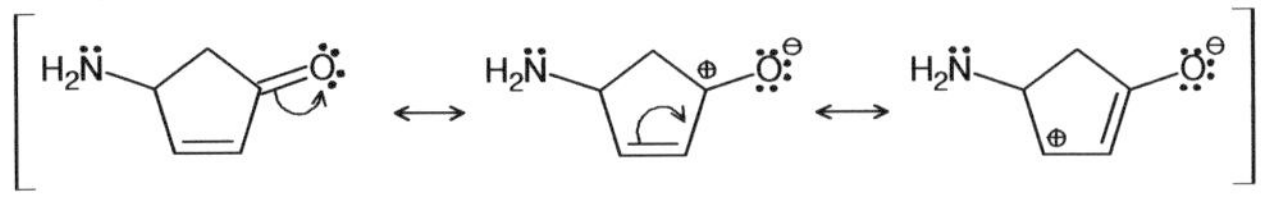

$H_2\ddot{N}$ $H_2\ddot{N}$ $H_2\ddot{N}$

Chapter 23

23.1

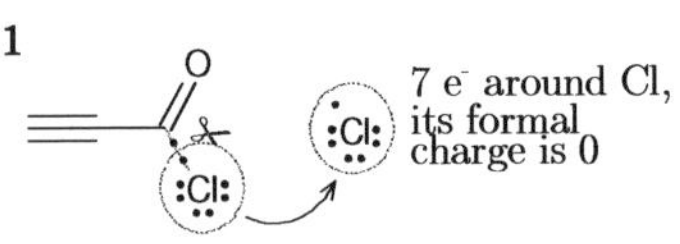

7 e⁻ around Cl, its formal charge is 0

23.2

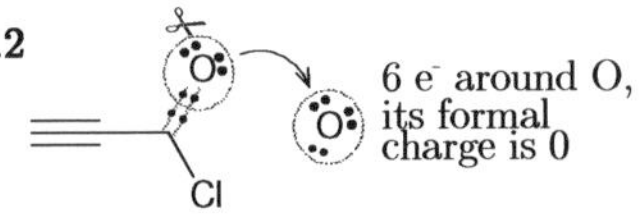

6 e⁻ around O, its formal charge is 0

23.3

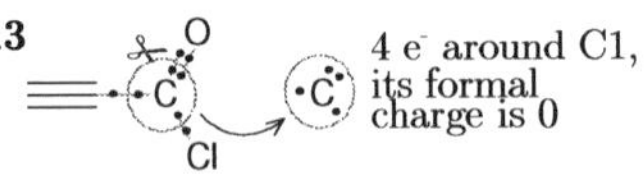

4 e⁻ around C1, its formal charge is 0

23.4

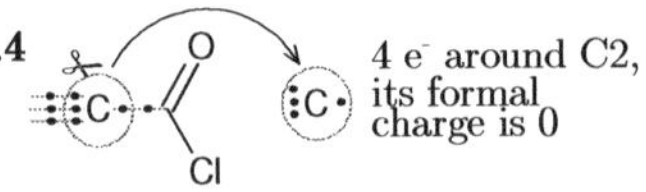

4 e⁻ around C2, its formal charge is 0

23.5

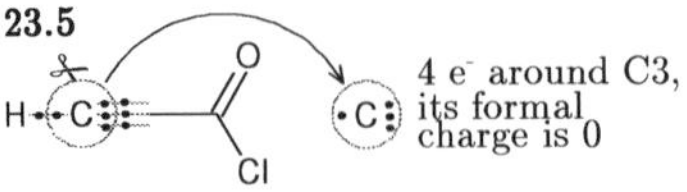

4 e⁻ around C3, its formal charge is 0

23.6

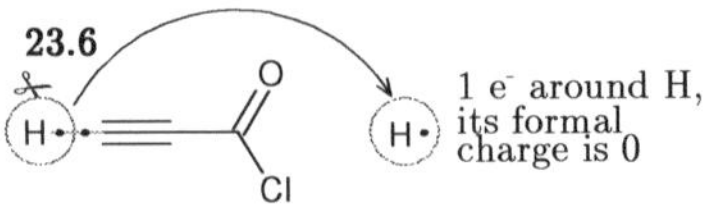

1 e⁻ around H, its formal charge is 0

Chapter 24

24.1

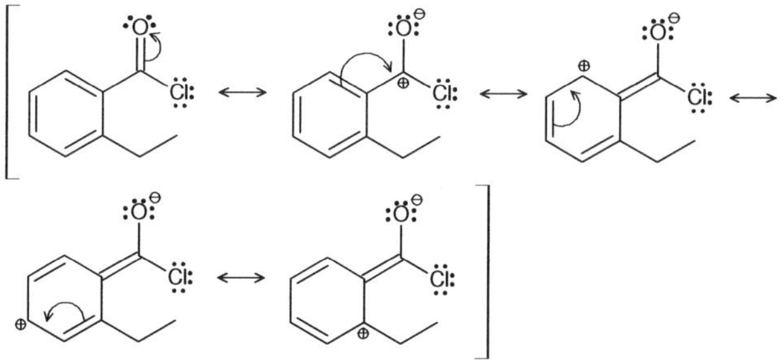

24.2

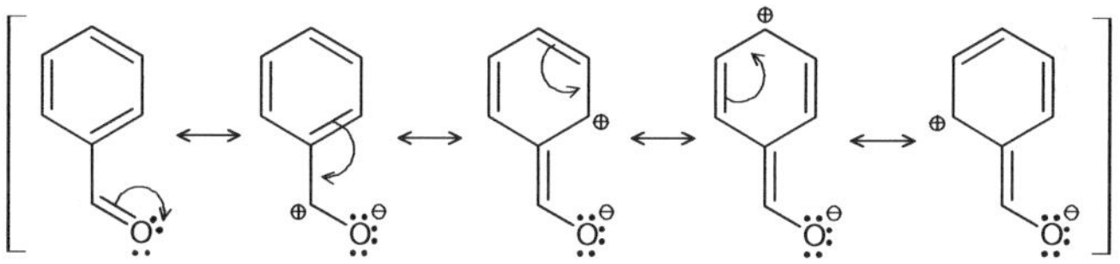

24.3

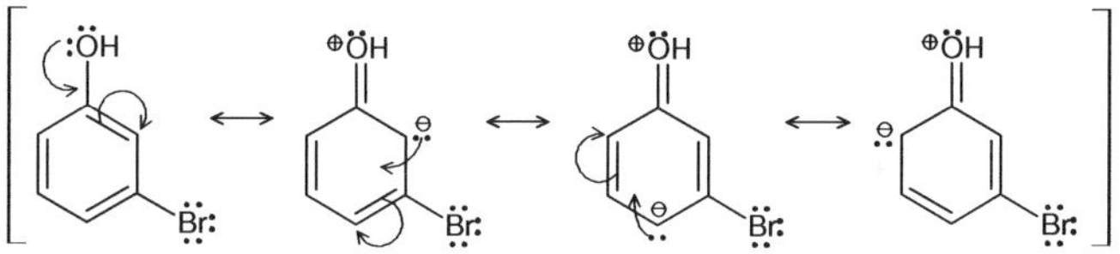

24.4

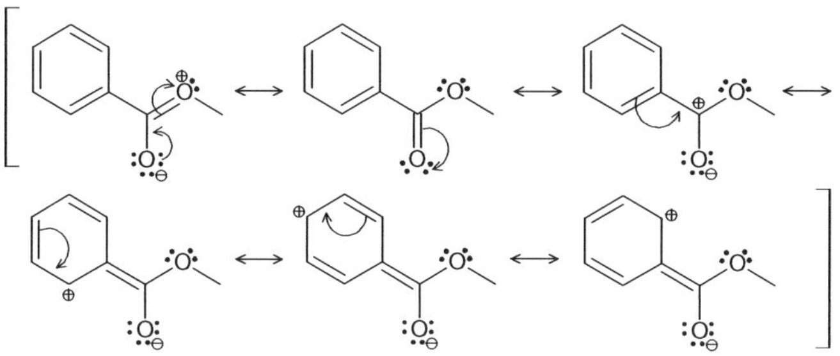

24.5

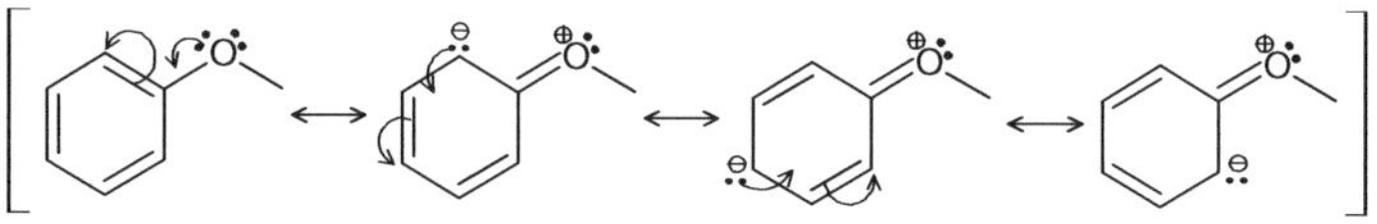

24.6

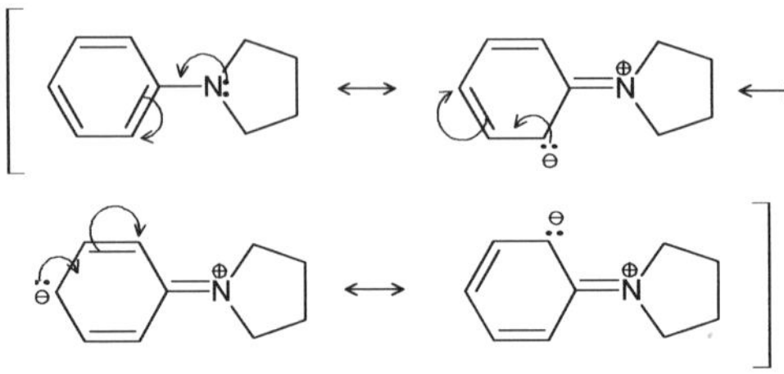

N, $N^{\oplus}$, $\ominus$

Chapter 27

27.1 $N^{\delta-} \leftarrow H^{\delta+}$ 27.2 $O^{\delta-} \leftarrow H^{\delta+}$ 27.3 $C^{\delta+} \rightarrow O^{\delta-}$ 27.4 $C^{\delta+} \Rightarrow O^{\delta-}$

27.5 $C^{\delta+} \rightarrow N^{\delta-}$ 27.6 $C^{\delta+} \Rightarrow N^{\delta-}$ 27.7 $C^{\delta+} \equiv N^{\delta-}$ 27.8 $C^{\delta+} \rightarrow Br^{\delta-}$

27.9

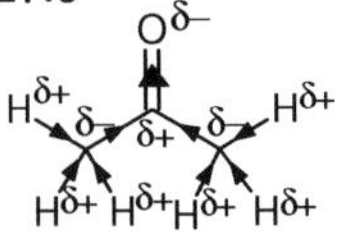

$O^{\delta-}$, $H^{\delta+}$, $\delta-$, $\delta+$, $\delta-$, $H^{\delta+}$, $H^{\delta+}$ $H^{\delta+}$ $H^{\delta+}$ $H^{\delta+}$

27.10

$H^{\delta+}$, $O^{\delta-}$, $\delta-$, $\delta+$, $H^{\delta+}$, $H^{\delta+}$, $O^{\delta-} \leftarrow H^{\delta+}$

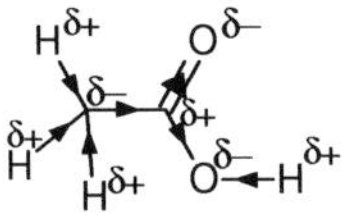

Chapter 28

28.1 [lowest] $E < D < C < A < B$ [highest]
28.2 [lowest] $B < A < C < D$ [highest]
28.3 [lowest] $A < C < B < D$ [highest]
28.4 [lowest] $D < C < B < A$ [highest]

Chapter 30

30.1

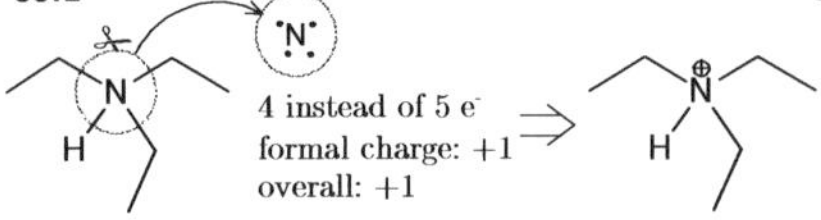

4 instead of 5 e⁻
formal charge: +1
overall: +1

30.2

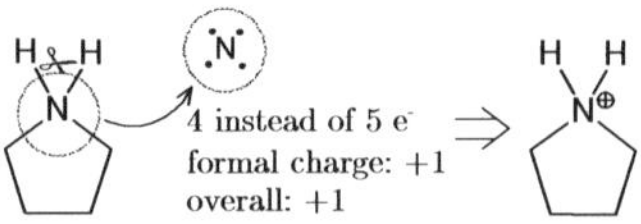

4 instead of 5 e⁻
formal charge: +1
overall: +1

30.3

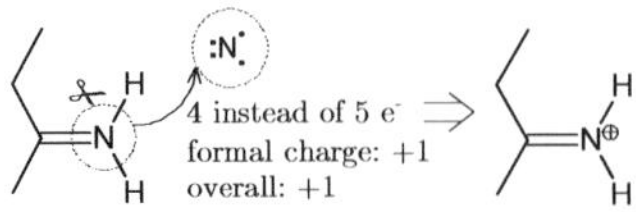

4 instead of 5 e⁻
formal charge: +1
overall: +1

30.4

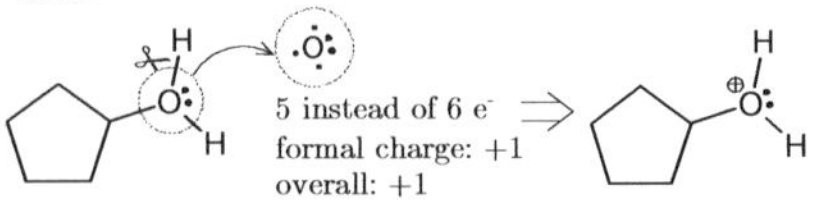

5 instead of 6 e⁻
formal charge: +1
overall: +1

30.5

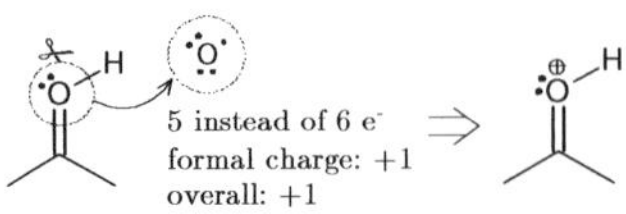

5 instead of 6 e⁻
formal charge: +1
overall: +1

30.6

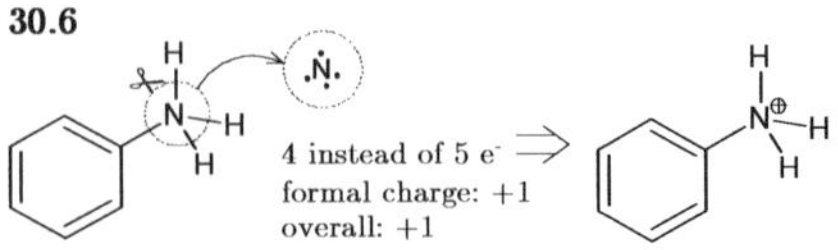

4 instead of 5 e⁻
formal charge: +1
overall: +1

30.7

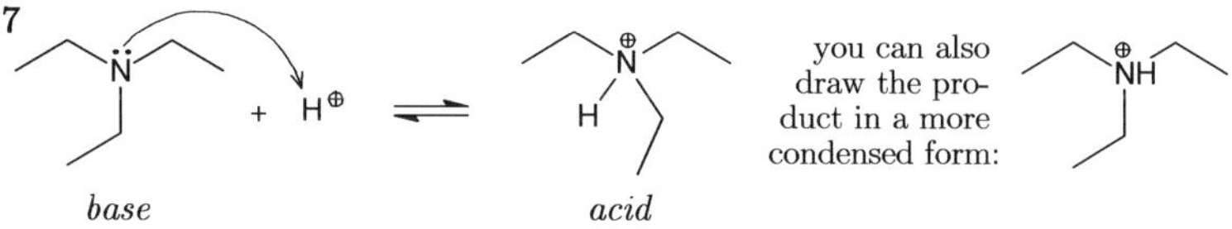

+ $H^{\oplus}$ ⇌

base *acid*

you can also draw the product in a more condensed form:

30.8

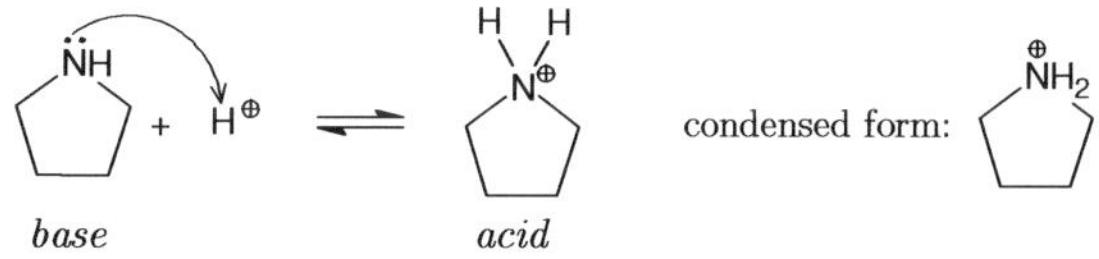

+ $H^{\oplus}$ ⇌

base *acid*

condensed form:

30.9

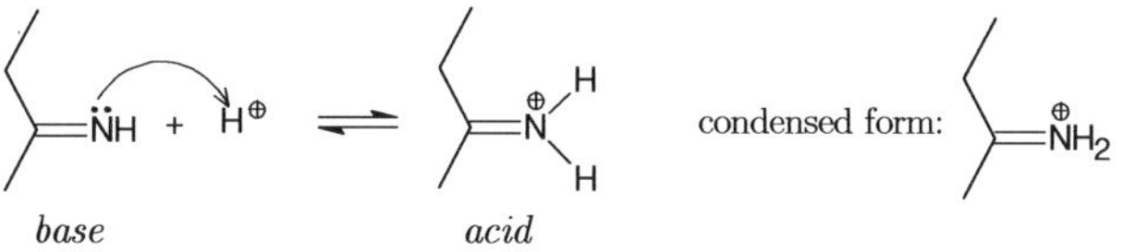

+ $H^{\oplus}$ ⇌

base *acid*

condensed form:

30.10

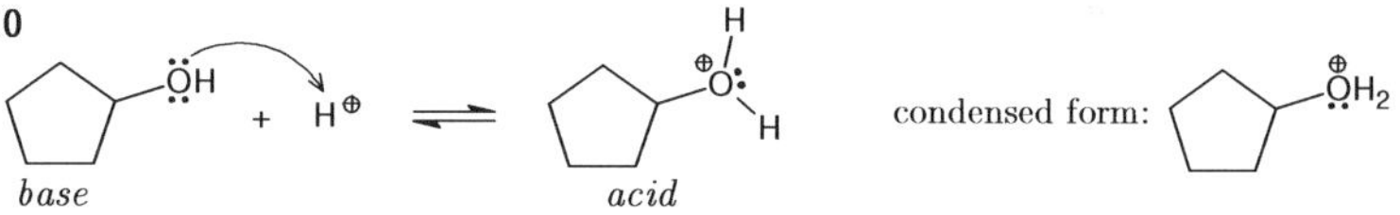

+ $H^{\oplus}$ ⇌

base *acid*

condensed form:

30.11

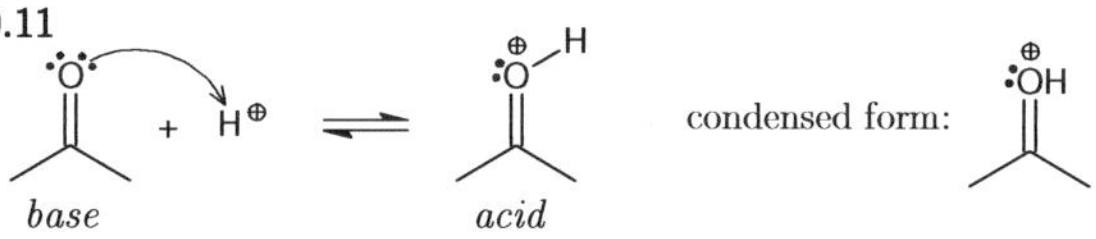

+ $H^{\oplus}$ ⇌

base *acid*

condensed form:

30.12

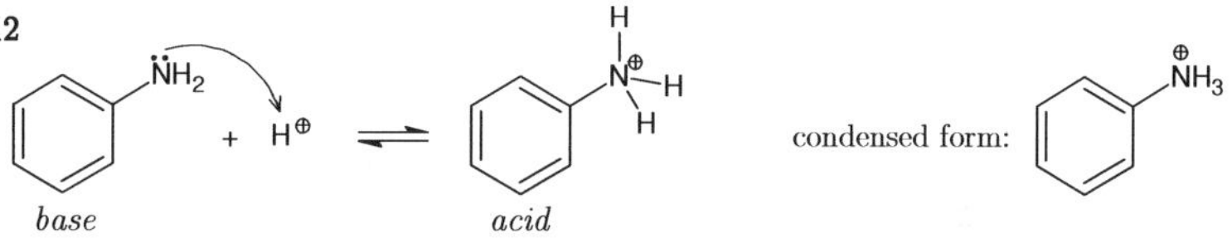

+ $H^{\oplus}$ ⇌

base *acid*

condensed form:

30.13

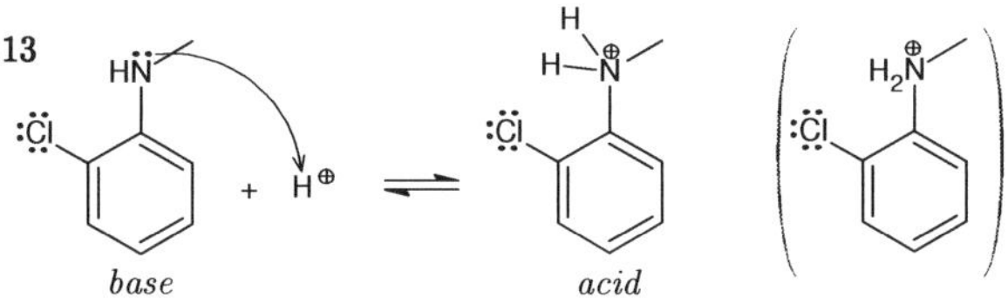

+ $H^{\oplus}$ ⇌

base *acid*

30.14

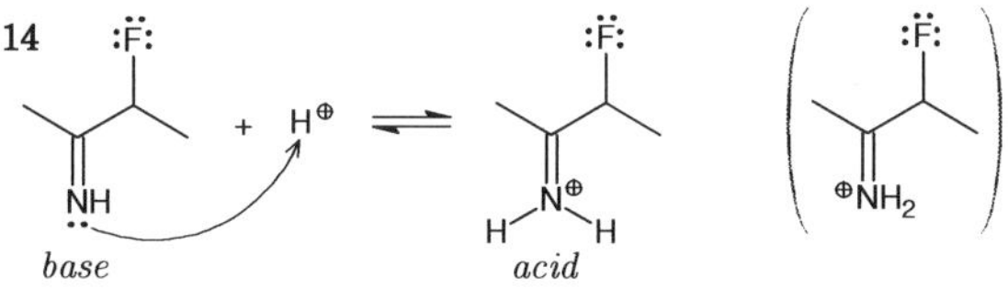

+ $H^{\oplus}$ ⇌

base *acid*

30.15

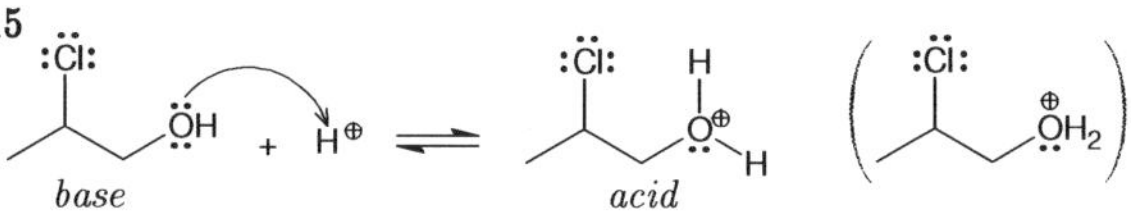

+ $H^{\oplus}$ ⇌

base *acid*

30.16

+ $H^{\oplus}$ ⇌

base *acid*

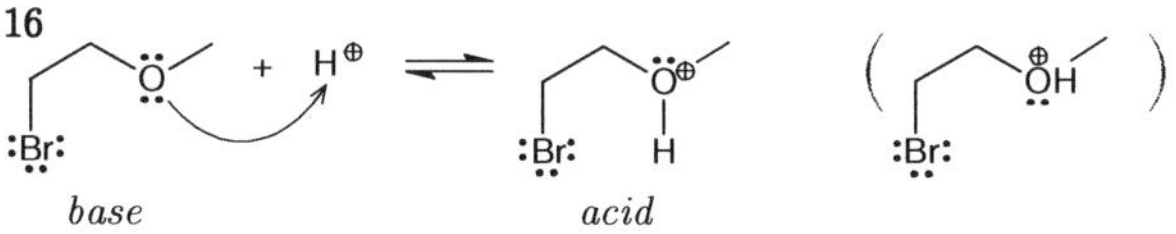

30.17

$H_2\ddot{N}$ $\ddot{O}H$ + $H^{\oplus}$ ⇌ $H-N^{\oplus}(H)(H)$ $\ddot{O}H$ ($H_3\overset{\oplus}{N}$ $\ddot{O}H$)

base *acid*

30.18

$H_2\ddot{N}$ $\ddot{O}$: + $H^{\oplus}$ ⇌ $H-N^{\oplus}(H)(H)$ $\ddot{O}$: ($H_3\overset{\oplus}{N}$ $\ddot{O}$:)

base *acid*

30.19

$\ddot{F}$: + $H^{\oplus}$ (crossed out arrows)

crossed out reaction arrows symbolize chemical reaction which does not occur

30.20

$\ddot{O}$: + $H^{\oplus}$ ⇌ $O^{\oplus}$–H ($\overset{\oplus}{O}H$)

base *acid*

30.21

+ $H^{\oplus}$ (crossed out arrows)

30.22 group A – nitrogen containing molecules (**30.13,14,17,18**)
group B – molecules with O, but no N atoms (**30.15,16,20**)
group C – molecules with no O and no N atoms (**30.19,21**)

Chapter 31

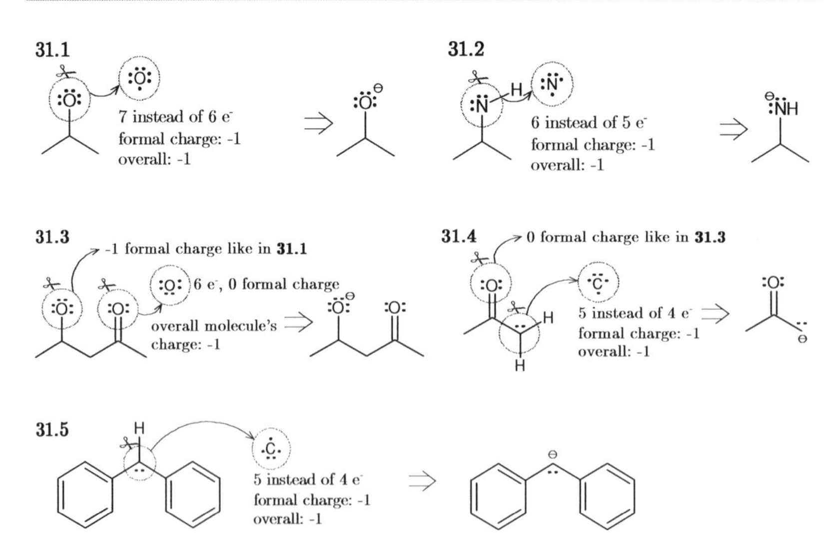

31.6

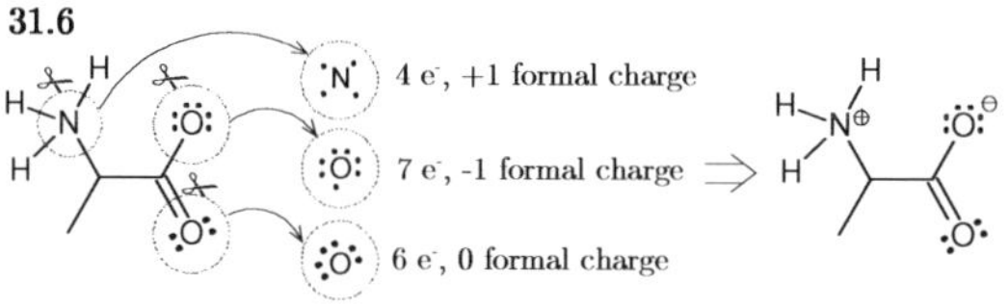

4 e⁻, +1 formal charge

7 e⁻, -1 formal charge

6 e⁻, 0 formal charge

Comment on 31.6: Overall charge of the molecule is 0. We refer to neutral molecules with one positive and one negative charge as **zwitterions**.

31.7

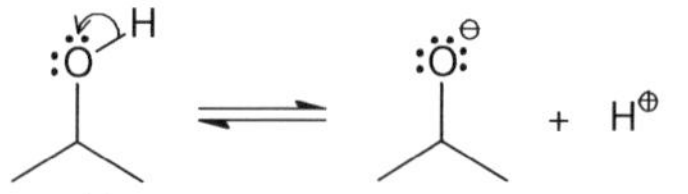

acid ⇌ *base* + $H^{\oplus}$

31.8

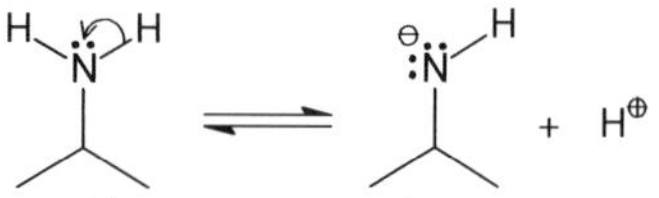

acid ⇌ *base* + $H^{\oplus}$

31.9

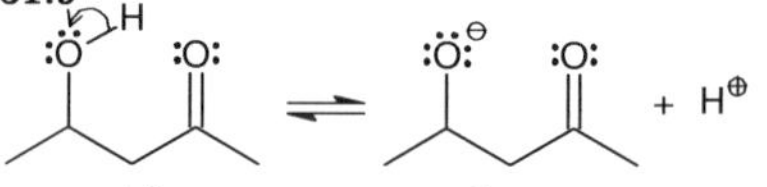

acid ⇌ *base* + $H^{\oplus}$

31.10

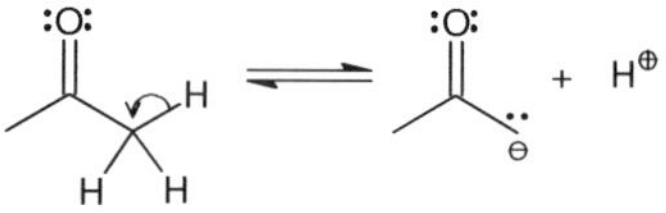

acid ⇌ *base* + $H^{\oplus}$

31.11

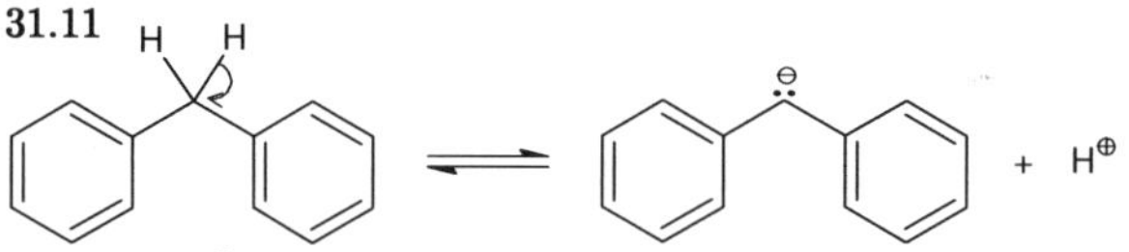

acid ⇌ *base* + $H^{\oplus}$

31.12

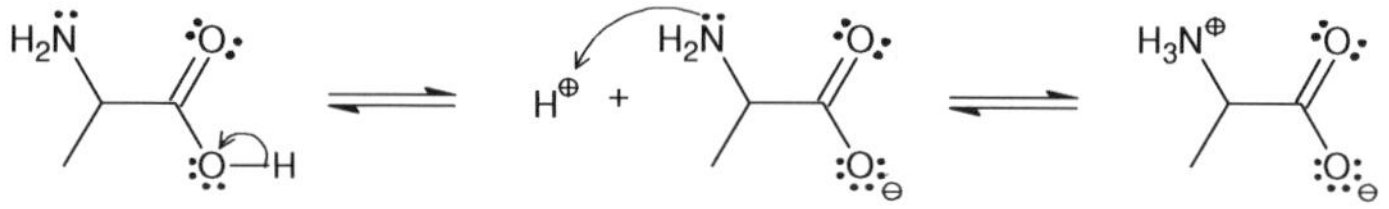

31.13

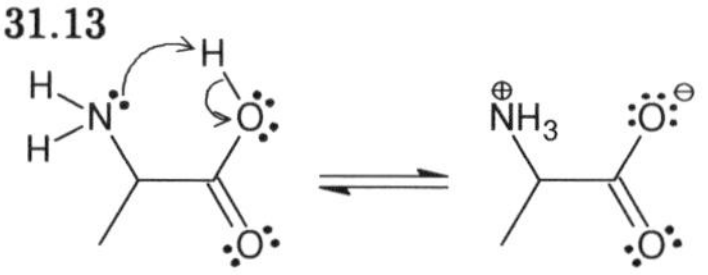

31.14

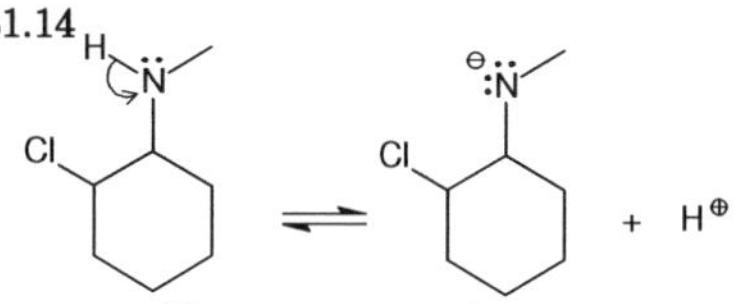

acid ⇌ *base* + $H^{\oplus}$

31.15

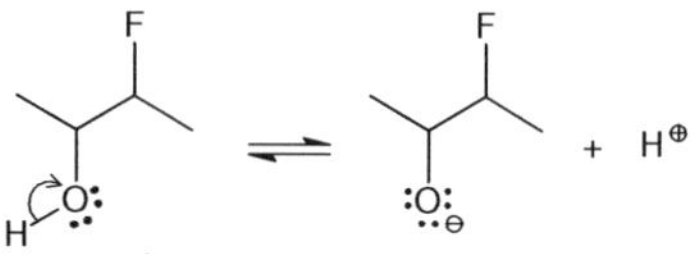

acid ⇌ *base* + $H^{\oplus}$

31.16

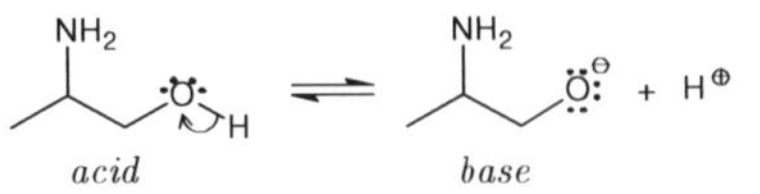

NH_2 ... O–H ⇌ NH_2 ... $O^{\ominus}$ + $H^{\oplus}$

acid *base*

31.17

NH_2 ... C(=O)–O–H ⇌ NH_2 ... C(=O)–$O^{\ominus}$ + $H^{\oplus}$

acid *base*

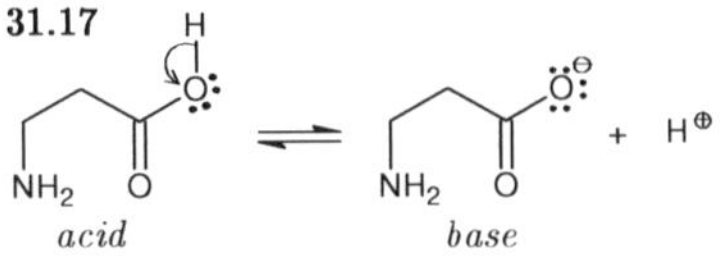

Chapter 32

32.1 The right amine is a stronger one, because it has a NH_2 group with localized lone pair, while in the left molecule the pair is delocalized.

32.2 The right amine is a stronger one, because it has a NH_2 group with localized lone pair, while in the left molecule the pair is delocalized.

32.3 Both amines are aromatic, so both lone pairs are delocalized. However, left one is a stronger base, thanks to an additional inductive effect from the methyl group. Since it is an electron donating one (N sucks its electron density), the electron density on N is higher.

32.4 The right amine is a stronger base, because in the left one a CH_2CH_2F group slightly withdraws electron density from NH_2. N is impoverished and holds its lone pair more tightly.

32.5 Electron density withdrawing effect of a CH_2CHF_2 group is larger than in CH_2CH_2F, thus the left amine is a stronger base.

32.6 The right amine is a stronger base, because of the inductive effect. The more branched the backbone the higher its electron donating ability.

32.7 The right imine is a stronger base, because of the inductive effect of an additional methyl group.

32.8 Right imine is a stronger base, because of the inductive effect of a larger carbon backbone.

32.9 Ethers are stronger bases than alcohols, because of the inductive effect. Both are hard to protonate, but in ethers electron density around O is higher, thanks to the presence of two alkyl groups.

32.10 In amide, N wins with O, but its lone pair is delocalized together with π pair of the C=O bond. Therefore, amine is much stronger base than amide.

32.11 Both amines are relatively weak bases, because their lone pairs are delocalized. Nevertheless, left one is stronger. In the right molecule, the inductive effect of F atom additionally decreases the electron density on the NH_2 group.

32.12 The left amine is stronger, because in the right molecule lone pair on N is delocalized together with a π pair.

32.13 The right structure is a stronger anionic base, because it experiences stronger electron donating influence from the alkyl group.

32.14 The right amine is a stronger base, because N experiences an inductive effect of three methyl groups, instead of just two.

32.15 The right structure is much stronger anionic base, because electron density around O atom is high (inductive effect of a large backbone). In the left anion, lone pair is delocalized around the benzene ring, and negative charge is distributed on four atoms.

32.16

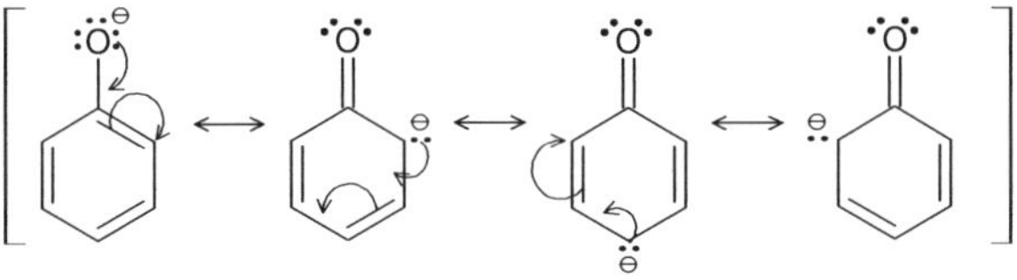

32.17 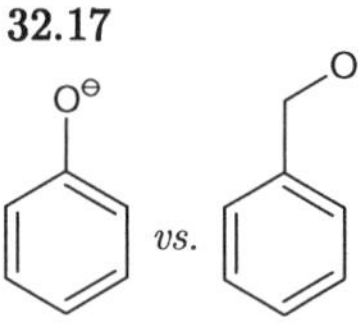 *vs.*

32.18 *vs.* F

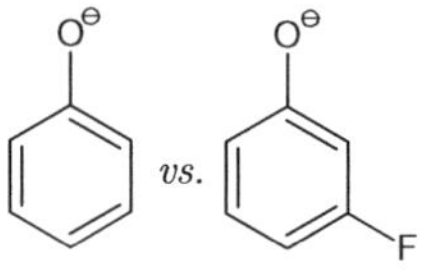

32.19 O NH *vs.* N

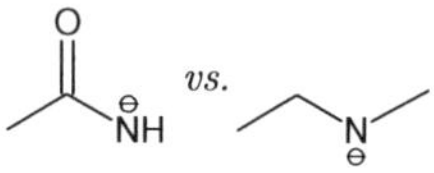

32.17 Phenol* (left) is a stronger acid than alcohol, because left conjugated base is weaker (lone pairs are delocalized; charge is distributed on four atoms). *We refer to aromatic alcohols (OH group bonds to the benzene ring directly), as **phenols**.

32.18 The right phenol is a stronger acid; because the right conjugated base is weaker (F atom withdraws electron density inductively).

32.19 Amides are stronger acids than amines. Here, the left conjugated base is indeed weaker, than the right one (lone pair is delocalized, and the charge is distributed on N and O atoms).

32.20 F HN *vs.* HN

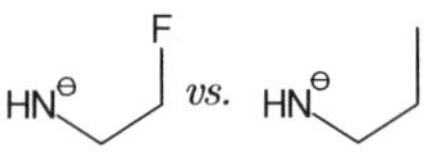

32.21 F O *vs.* F O

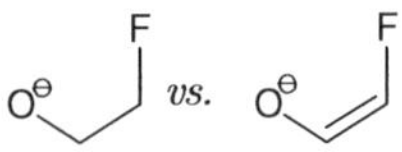

32.22 O *vs.* O

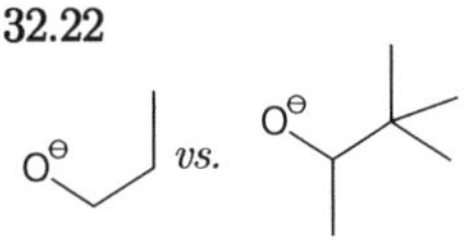

32.20 Amines are very weak acids, but in harsh conditions, it is possible to deprotonate the NH_2 group. The left amine is a slightly stronger acid, because the left conjugate base is weaker (due to the inductive effect of F).

32.21 Right alcohol is a stronger acid, because right conjugated base is weaker (lone pair is delocalized, and charge is distributed on O and one C atom).

32.22 Left alcohol is a stronger acid; because left conjugated base is weaker (experiences smaller influence of alkyls, which donate electron density).

32.23 $O^{\ominus}$ *vs.* $O^{\ominus}$ O O

32.24 $O^{\ominus}$ *vs.* $O^{\ominus}$ Cl O O

32.25 $O^{\ominus}$ *vs.* $O^{\ominus}$ O O

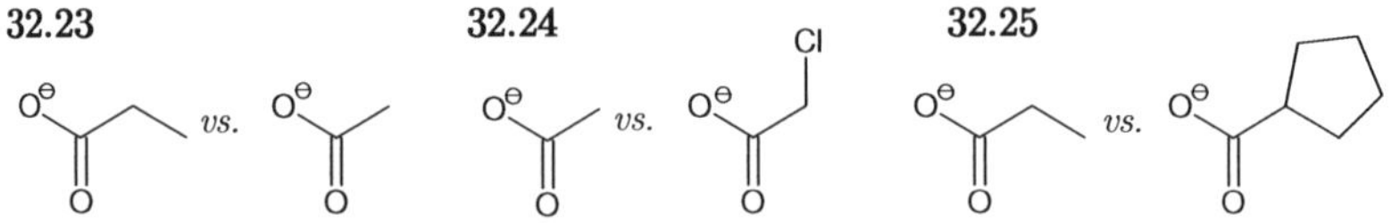

32.23 Both molecules are good acids, but ethanoic acid is stronger, because its conjugate base is weaker (conjugated base of propanoic acid experiences higher inductive influence of electron donating chain).

32.24 Right acid is stronger, because right conjugate base is weaker (inductive influence of Cl).

32.25 Left acid is stronger, because left conjugated base is weaker (the inductive effect of electron donating alkyl group is weaker, than in the right structure).

32.26 NH_2 *vs.* NH_2

32.27 NH_2 *vs.* NH

32.28 NH_2 *vs.* NH_2

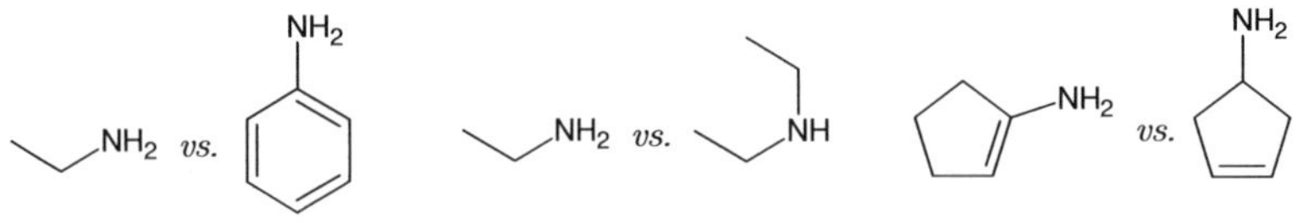

32.26 The right structure is a stronger acid; because the right conjugated amine is weaker (its lone pair is delocalized).

32.27 The left structure is a stronger acid, because the left conjugated amine is weaker.

32.28 The left structure is a stronger acid, because the left conjugated amine is weaker.

Chapter 33

33.1

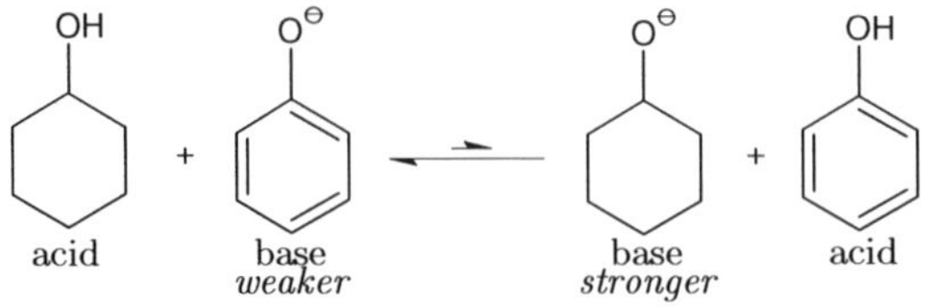

OH + $O^{\ominus}$ ⇌ $O^{\ominus}$ + OH

acid | base *weaker* | base *stronger* | acid

33.2

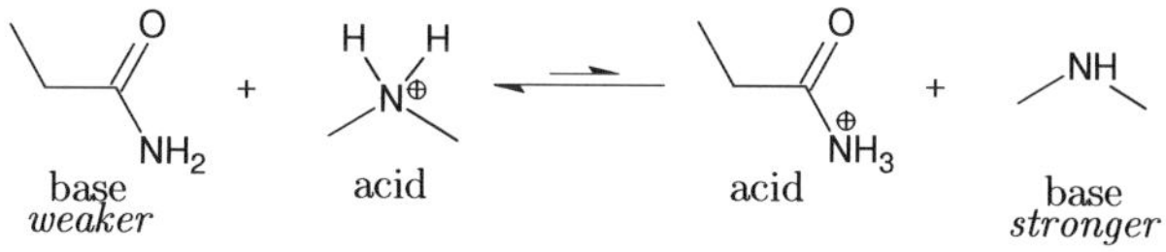

O, NH_2 + H H $N^{\oplus}$ ⇌ O, $NH_3^{\oplus}$ + NH

base *weaker* | acid | acid | base *stronger*

33.3

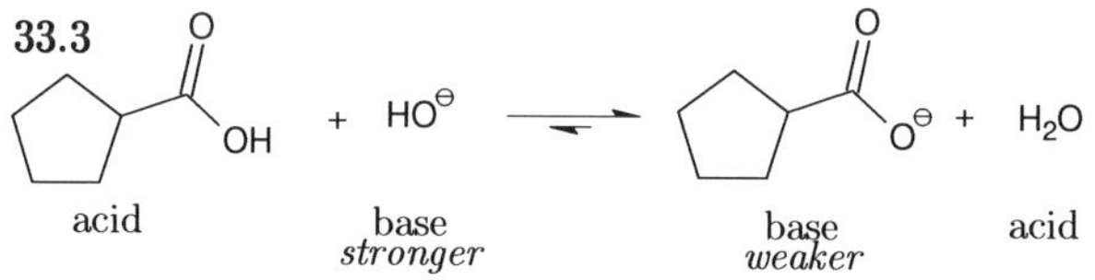

O, OH + $HO^{\ominus}$ ⇌ O, $O^{\ominus}$ + H_2O

acid | base *stronger* | base *weaker* | acid

33.4

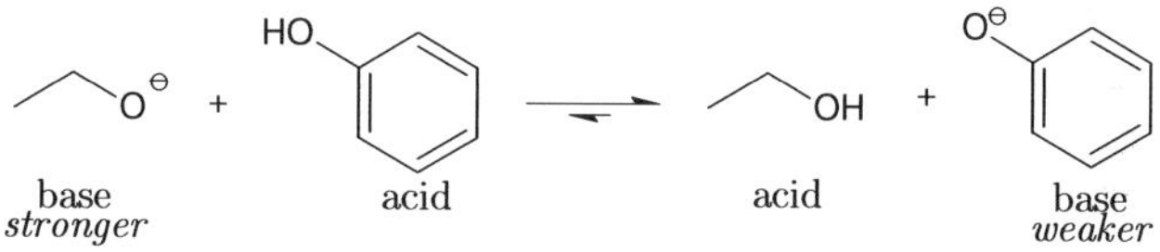

$O^{\ominus}$ + HO ⇌ OH + $O^{\ominus}$

base *stronger* | acid | acid | base *weaker*

33.5

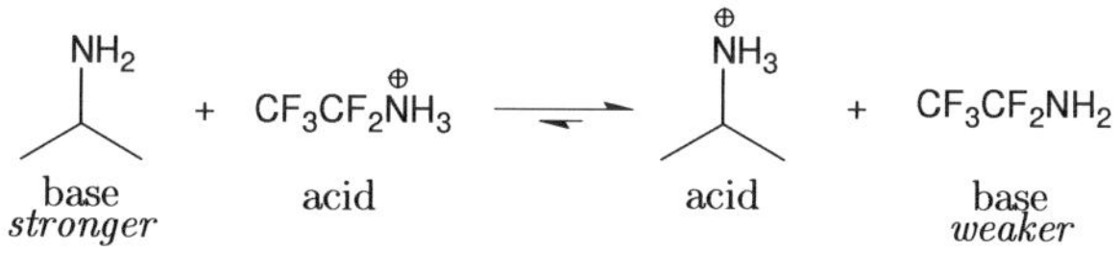

NH_2 + $CF_3CF_2\overset{\oplus}{N}H_3$ ⇌ $\overset{\oplus}{N}H_3$ + $CF_3CF_2NH_2$

base *stronger* | acid | acid | base *weaker*

33.6

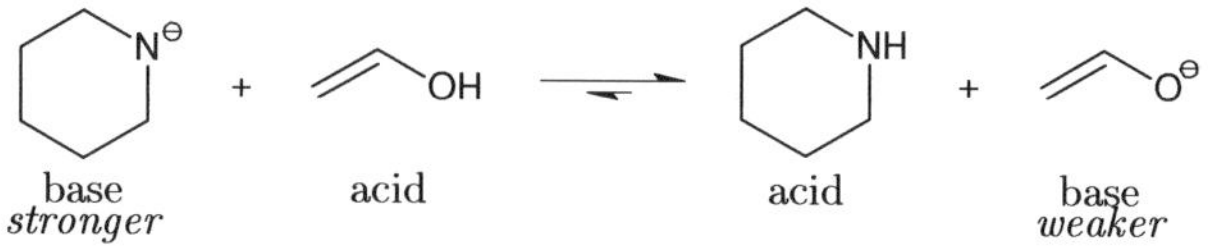

$N^{\ominus}$ + OH ⇌ NH + $O^{\ominus}$

base *stronger* | acid | acid | base *weaker*

33.7

$O^{\ominus}$ + H_2O ⇌ OH + $HO^{\ominus}$

base *stronger* | acid | acid | base *weaker*

33.8

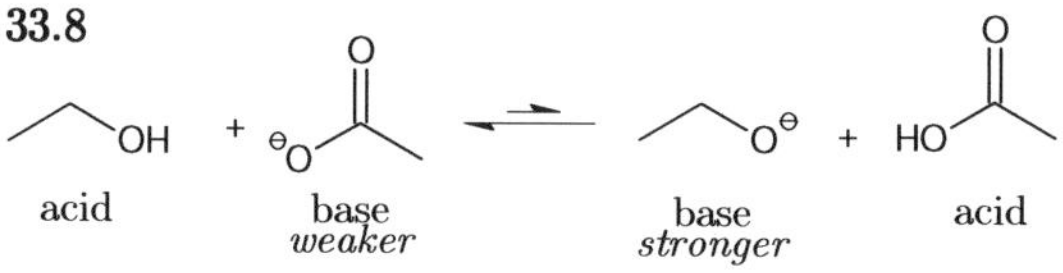

OH + O, $^{\ominus}O$ ⇌ $O^{\ominus}$ + O, HO

acid | base *weaker* | base *stronger* | acid

33.9

$$O^{\ominus} + NH \rightleftharpoons OH + N^{\ominus}$$

base *weaker* | acid | acid | base *stronger*

33.10

$$H_3O^{\oplus} + NH \rightleftharpoons H_2O + NH_2^{\oplus}$$

acid | base *stronger* | base *weaker* | acid

Index

Printed in Great Britain
by Amazon